ESSENTIALS OF GEOLOGY

Edited by
Judith Rosales

Delve Publishing

Essentials of Geology

Edited by Judith Rosales

Edition 2018

Delve Publishing

Published by
Delve Publishing LLC
708 3rd Avenue, 6th Floor
New York
NY 10801
United States of America
www.delvepublishing.com

ISBN: 978-1-68095-735-8 (Hardback)

Printed in United States of America

Library of Congress Control Number: 2016920415

About the Editor

Judith Rosales

Prof. Judith is an Ecologist and Environmental Specialist with more than 30 years of academic and consulting experience. She holds a PhD in geography from the University of Birmingham UK and wrote the PhD and Master Academic Programs of Environmental Sciences for the Universidad de Guayana, Venezuela.

Table of Contents

2 HISTORY OF GEOLOGY

3 PRINCIPLES OF GEOLOGY

4 GEOLOGIC PROVINCE

5 PLATE TECTONICS

6 STRUCTURAL GEOLOGY

7 PLANETARY GEOLOGY AND GEOLOGY OF SOLAR TERRESTRIAL PLANETS

8 ENGINEERING GEOLOGY AND GEOTECHNICAL ENGINEERING

9 ECONOMIC GEOLOGY

List of Abbreviations

CCS	Carbon capture and storage
CO	Carbon monoxide
CAMP	Central Atlantic Magmatic Province
CTL	Coal to Liquid
CWS	Coal-water slurry fuel
IGCC	Impurities from the coal, gasification
ICE	Intercontinental Exchange
IOCG	Iron oxide copper gold
LAPs	Large andesitic provinces
LBRPs	Large basaltic–rhyolitic provinces
LGP	Large granitic provinces
LIP	Large igneous province
LPP	Large plutonic provinces
LRPs	Large rhyolitic provinces
LVP	Large volcanic provinces
GIS	Geographic Information System
GPS	Global Positioning System
GPR	Ground penetrating radar
NRM	Natural remanent magnetization
PGE	Platinum-Group Element
SDRS	Seaward-dipping reflector sequences
UHP	Ultra-High-Pressure
TNOs	Trans-Neptunian objects
STD	Submarine tailings disposal

Preface

Geology means, literally, the study of the earth, the materials of which it is made, the structure of those materials, and the processes acting upon them. It includes the study of organisms that have inhabited our planet. An important part of geology is the study of how Earth's materials, structures, processes and organisms have changed over time. Geologists work to understand the history of our planet. The better they can understand Earth's history the better they can foresee how events and processes of the past might influence the future. Geology is also about understanding the evolution of life on Earth; about discovering resources such as metals and energy; about recognizing and minimizing the environmental implications of our use of those resources; and about learning how to mitigate the hazards related to earthquakes, volcanic eruptions, and slope failures. All of these aspects of geology, and many more, are covered in this textbook. Geology gives insight into the history of the Earth by providing the primary evidence for plate tectonics, the evolutionary history of life, and past climates. Geology is important for mineral and hydrocarbon exploration and exploitation, evaluating water resources, understanding of natural hazards, the remediation of environmental problems, and for providing insights into past climate change. The majority of geological data comes from research on solid Earth materials. These typically fall into one of two categories: rock and unconsolidated material. Seismologists can use the arrival times of seismic waves in reverse to image the interior of the Earth. Early advances in this field showed the existence of a liquid outer and a dense solid inner core. Methods for relative dating were developed when geology first emerged as a formal science. Geologists still use the following principles today as a means to provide information about geologic history and the timing of geologic events. Geology also plays a role in geotechnical engineering and is a major academic discipline.

Content Coverage

Chapters One and *Two* focus on geology fundamentals and history of the geology, respectively. The history of geology is concerned with the development of the natural science of geology.

Chapters Three and Four are about principles of geology and geologic province, separately. A province may include a single dominant structural element such as a basin or a fold belt, or a number of contiguous related elements.

Chapter Five provides the theory of plate tectonics which deals with the dynamics of earth's outer shell, the lithosphere that revolutionized Earth sciences by providing a uniform context for understanding mountain-building processes, volcanoes, and earthquakes, as well as understanding the evolution of Earth's surface and reconstructing its past continental and oceanic configurations.

Chapters Six and *Seven* present information about structural geology and planetary geology, separately. Structural geology focuses on the study of geological structures, with the goal of learning how, when, and why they formed, whereas planetary geology is concerned with the geology of the celestial bodies such as the planets and their moons, asteroids, comets, and meteorites

Chapters Eight and *Nine* present information about engineering geology and economic geology, seperately. Engineering geology is concerned with how the geological factors regarding the location, design, construction, operation and maintenance of engineering works are recognized and accounted for, whereas Economic geology deals with metal ores, fossil fuels, and other materials of commercial value, such as salt, gypsum, and building stone.

CHAPTER 1

Introduction to Geology

Chapter Contents

- Introduction
- Geologic Materials
- Whole-Earth Structure
- Geologic Time
- Dating Methods
- Geological Development of an Area
- Methods of Geology
- Planetary Geology
- Applied Geology
- History of Geology

Geology means, literally, the study of the Earth. Explore this section to understand the structure of the Earth and its surface features, what causes earthquakes and tsunamis, and why volcanoes form and erupt. Learn about minerals, which form the building blocks of rocks, and how rocks are made and destroyed. Learn about Earth's fascinating history, the variety of life forms which have roamed the surface over the millennia, and the dramatic changes that have happened over Earth's long history.

INTRODUCTION

Geology is an earth science comprising the study of solid Earth, the rocks of which it is composed, and the processes by which they change. Geology can also refer generally to the study of the solid features of any celestial body (such as the geology of the Moon or Mars). Geology gives insight into the history of the Earth by providing the primary evidence for plate tectonics, the evolutionary history of life, and past climates. Geology is important for mineral and hydrocarbon exploration and exploitation, evaluating water resources, understanding of natural hazards, the remediation of environmental problems, and for providing insights into past climate change. Geology also plays a role in geotechnical engineering and is a major academic discipline.

GEOLOGIC MATERIALS

The majority of geological data comes from research on solid Earth materials. These typically fall into one of two categories: rock and unconsolidated material.

Rock

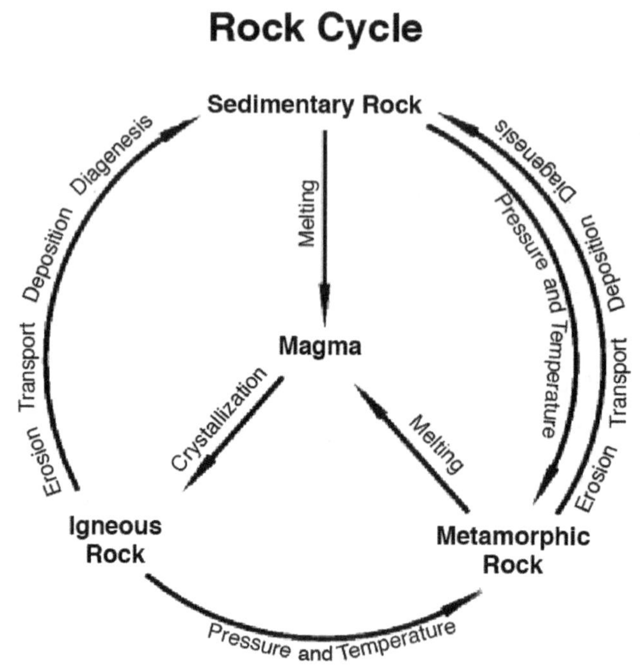

This schematic diagram of the rock cycle shows the relationship between magma and sedimentary, metamorphic, and igneous rock

There are three major types of rock: igneous, sedimentary, and metamorphic. The rock cycle is an important concept in geology which illustrates the relationships between these three types of rock, and magma. When a rock crystallizes from melt (magma and/or lava), it is an igneous rock. This rock can be weathered and eroded, and then redeposited and lithified into a sedimentary rock, or be turned into a metamorphic rock due to heat and pressure that change the mineral content of the rock which gives it a characteristic fabric. The sedimentary rock can then be subsequently turned into a metamorphic rock due to heat and pressure and is then weathered, eroded, deposited, and lithified, ultimately becoming a sedimentary rock. Sedimentary rock may also be re-eroded and redeposited, and metamorphic rock may also undergo additional metamorphism. All three types of rocks may be re-melted; when this happens, a new magma is formed, from which an igneous rock may once again crystallize.

The majority of research in geology is associated with the study of rock, as rock provides the primary record of the majority of the geologic history of the Earth.

Unconsolidated Material

Geologists also study unlithified material, which typically comes from more recent deposits. These materials are superficial deposits which lie above the bedrock. Because of this, the study of such material is often known as Quaternary geology, after the recent Quaternary Period. This includes the study of sediment and soils, including studies in geomorphology, sedimentology, and paleoclimatology.

WHOLE-EARTH STRUCTURE

Plate tectonics

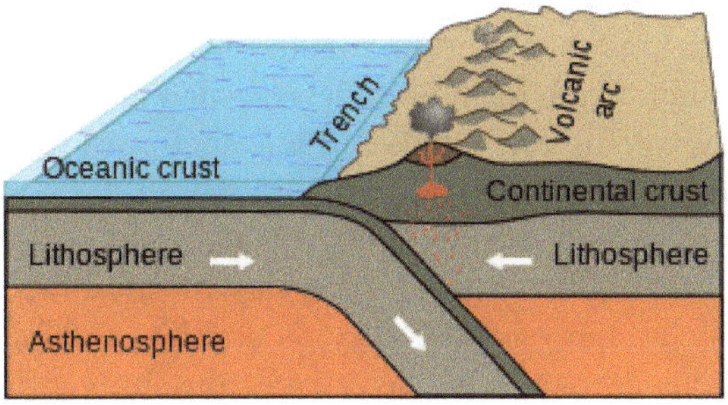

Oceanic-continental convergence resulting in subduction and volcanic arcs illustrates one effect of plate tectonics.

In the 1960s, it was discovered that the Earth's lithosphere, including the crust and rigid uppermost portion of the upper mantle, is separated into a number of tectonic plates that move across the plastically deforming, solid, upper mantle, which is called the asthenosphere. This theory is supported by a number of observations, including seafloor spreading, and the global distribution of mountain terrain and seismicity.

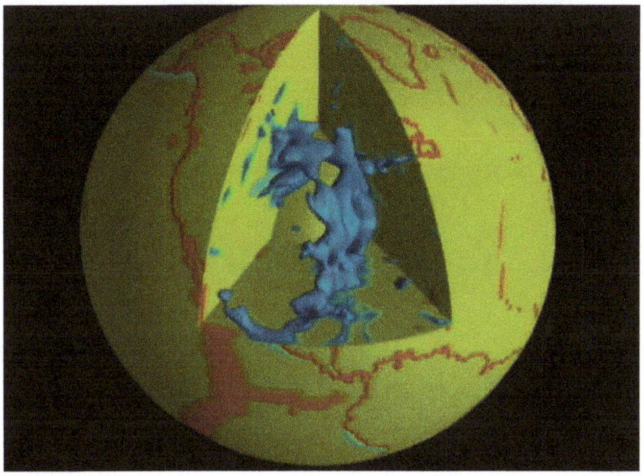

There is an intimate coupling between the movement of the plates on the surface and the convection of the mantle: oceanic plate motions and mantle convection currents always move in the same direction, because the oceanic lithosphere is the rigid upper thermal boundary layer of the convecting mantle. This coupling between rigid plates moving on the surface of the Earth and the convecting mantle is called plate tectonics.

On this diagram, subducting slabs are in blue, and continental margins and a few plate boundaries are in red. The blue blob in the cutaway section is the seismically imaged Farallon Plate, which is subducting beneath North America. The remnants of this plate on the Surface of the Earth are the Juan de Fuca Plate and Explorer plate in the Northwestern USA / Southwestern Canada, and the Cocos Plate on the west coast of Mexico.

The development of plate tectonics provided a physical basis for many observations of the solid Earth. Long linear regions of geologic features could be explained as plate boundaries. Mid-ocean ridges, high regions on the seafloor where hydrothermal vents and volcanoes exist, were explained as divergent boundaries, where two plates move apart. Arcs of volcanoes and earthquakes were explained as convergent boundaries, where one plate subducts under another. Transform boundaries, such as the San Andreas Fault system, resulted in widespread powerful earthquakes. Plate tectonics also provided a mechanism for Alfred Wegener's theory of continental drift, in which the continents move across the surface of the Earth over geologic time. They also provided a driving force for crustal deformation, and a new setting for the observations of structural geology. The power of the theory of plate tectonics lies in its ability to combine all of these observations into a single theory of how the lithosphere moves over the convecting mantle.

Earth Structure

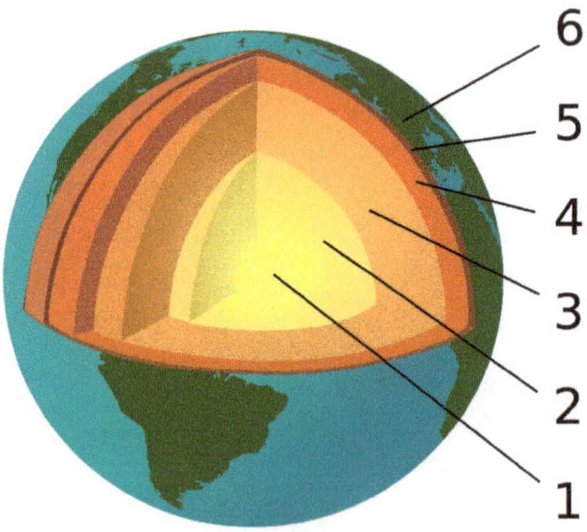

The Earth's layered structure. (1) inner core; (2) outer core; (3) lower mantle; (4) upper mantle; (5) lithosphere; (6) crust (part of the lithosphere)

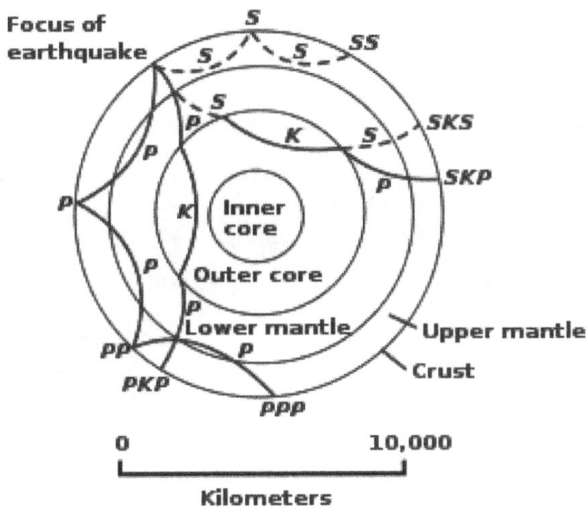

Earth layered structure. Typical wave paths from earthquakes like these gave early seismologists insights into the layered structure of the Earth

Advances in seismology, computer modeling, and mineralogy and crystallography at high temperatures and pressures give insights into the internal composition and structure of the Earth.

Seismologists can use the arrival times of seismic waves in reverse to image the interior of the Earth. Early advances in this field showed the existence of a liquid outer core (where shear waves were not able to propagate) and a dense solid inner core. These advances led to the development of a layered model of the Earth, with a crust and lithosphere on top, the mantle below (separated within itself by seismic discontinuities at 410 and 660 kilometers), and the outer core and inner core below that. More recently, seismologists have been able to create detailed images of wave speeds inside the earth in the same way a doctor images a body in a CT scan. These images have led to a much more detailed view of the interior of the Earth, and have replaced the simplified layered model with a much more dynamic model. Mineralogists have been able to use the pressure and temperature data from the seismic and modelling studies alongside knowledge of the elemental composition of the Earth to reproduce these conditions in experimental settings and measure changes in crystal structure. These studies explain the chemical changes associated with the major seismic discontinuities in the mantle and show the crystallographic structures expected in the inner core of the Earth.

GEOLOGIC TIME

The geologic time scale encompasses the history of the Earth. It is bracketed at the old end by the dates of the earliest Solar System material at 4.567 Ga, (gigaannum: billion years ago) and the age of the Earth at 4.54 Ga at the beginning of the informally recognized Hadean eon. At the young end of the scale, it is bracketed by the present day in the Holocene epoch.

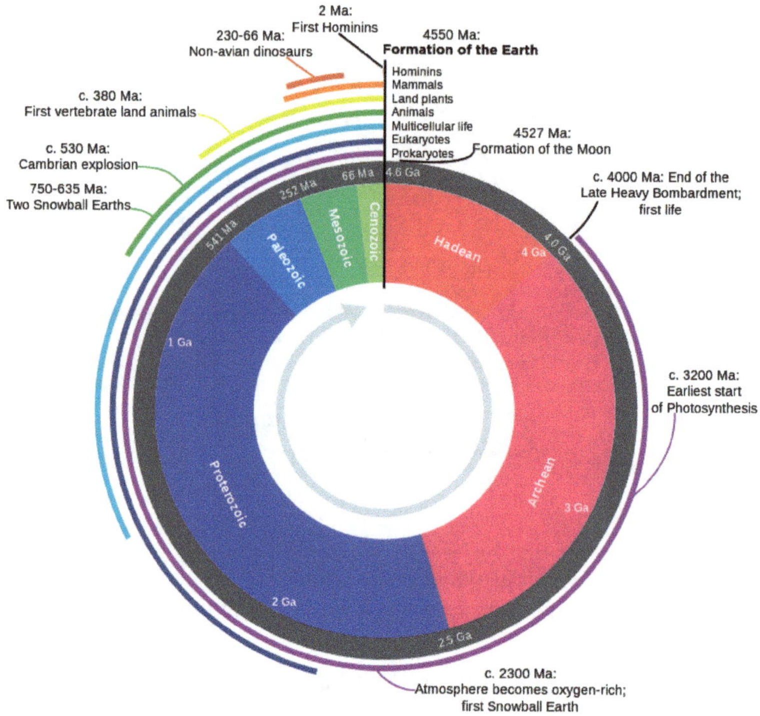

Geological time put in a diagram called a geological clock, showing the relative lengths of the eons of the Earth's history.

Important Milestones

- 4.567 Ga: Solar system formation
- 4.54 Ga: Accretion of Earth
- c. 4 Ga: End of Late Heavy Bombardment, first life
- c. 3.5 Ga: Start of photosynthesis
- c. 2.3 Ga: Oxygenated atmosphere, first snowball Earth
- 730–635 Ma (megaannum: million years ago): second snowball Earth
- 542 ± 0.3 Ma: Cambrian explosion – vast multiplication of hard-bodied life; first abundant fossils; start of the Paleozoic
- c. 380 Ma: First vertebrate land animals
- 250 Ma: Permian-Triassic extinction – 90% of all land animals die; end of Paleozoic and beginning of Mesozoic
- 66 Ma: Cretaceous–Paleogene extinction – Dinosaurs die; end of Mesozoic and beginning of Cenozoic
- c. 7 Ma: First hominins appear
- 3.9 Ma: First Australopithecus, direct ancestor to modern Homo sapiens, appear

- 200 ka (kiloannum: thousand years ago): First modern Homo sapiens appear in East Africa

Brief Time Scale

The following four timelines show the geologic time scale. The first shows the entire time from the formation of the Earth to the present, but this compresses the most recent eon. Therefore, the second scale shows the most recent eon with an expanded scale. The second scale compresses the most recent era, so the most recent era is expanded in the third scale. The third scale compresses the most recent period, so the most recent period is expanded in the fourth scale.

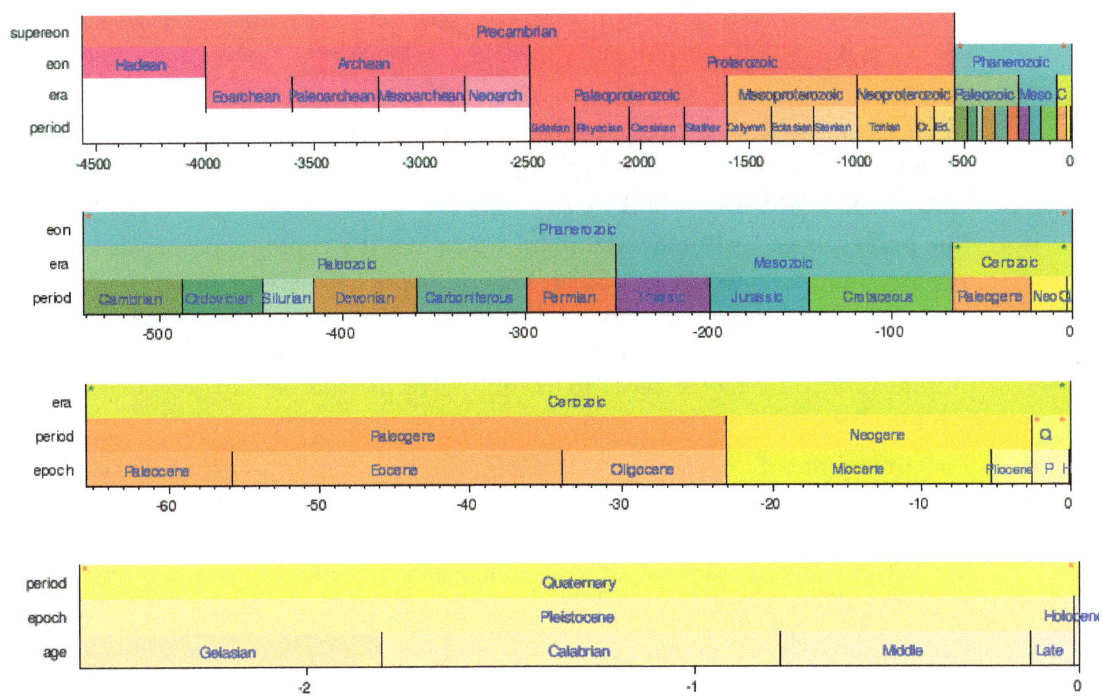

Millions of Years

DATING METHODS

Relative Dating

Cross-cutting relations can be used to determine the relative ages of rock strata and other geological structures. Explanations: A – folded rock strata cut by a thrust fault; B – large intrusion (cutting through A); C – erosional angular unconformity (cutting off A & B) on which rock strata were deposited; D – volcanic dyke (cutting through A, B & C); E – even younger rock strata (overlying C & D); F – normal fault (cutting through A, B, C & E).

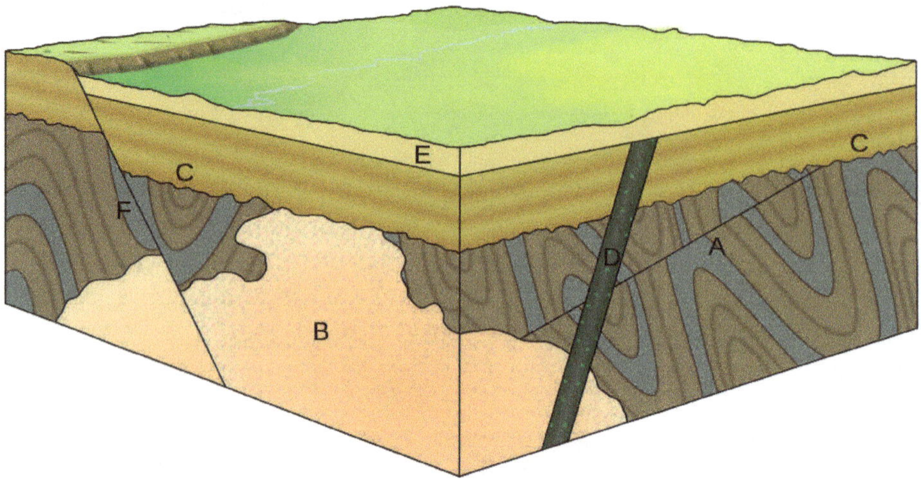

Methods for relative dating were developed when geology first emerged as a formal science. Geologists still use the following principles today as a means to provide information about geologic history and the timing of geologic events.

- **The principle of Uniformitarianism** states that the geologic processes observed in operation that modify the Earth's crust at present have worked in much the same way over geologic time. A fundamental principle of geology advanced by the 18th century Scottish physician and geologist James Hutton, is that the present is the key to the past. In Hutton's words: the past history of our globe must be explained by what can be seen to be happening now.
- **The principle of intrusive relationships** concerns crosscutting intrusions. In geology, when an igneous intrusion cuts across a formation of sedimentary rock, it can be determined that the igneous intrusion is younger than the sedimentary rock. There are a number of different types of intrusions, including stocks, laccoliths, batholiths, sills and dikes.
- **The principle of cross-cutting relationships** pertains to the formation of faults and the age of the sequences through which they cut. Faults are younger than the rocks they cut; accordingly, if a fault is found that penetrates some formations but not those on top of it, then the formations that were cut are older than the fault, and the ones that are not cut must be younger than the fault. Finding the key bed in these situations may help determine whether the fault is a normal fault or a thrust fault.
- **The principle of inclusions and components** states that, with sedimentary rocks, if inclusions (or *clasts*) are found in a formation, then the inclusions must be older than the formation that contains them. For example, in sedimentary rocks, it is common for gravel from an older formation to be ripped up and included in a newer layer. A similar situation with igneous rocks occurs when xenoliths are found. These foreign bodies are picked up as magma or lava flows, and are incorporated, later to cool in the matrix. As a result, xenoliths are older than the rock which contains them.

The Permian through Jurassic stratigraphy of the Colorado Plateau area of southeastern Utah is an example of both original horizontality and the law of superposition. These strata make up much of the famous prominent rock formations in widely spaced protected areas such as Capitol Reef National Park and Canyonlands National Park. From top to bottom: Rounded tan domes of the Navajo Sandstone, layered red Kayenta Formation, cliff-forming, vertically jointed, red Wingate Sandstone, slope-forming, purplish Chinle Formation, layered, lighter-red Moenkopi Formation, and white, layered Cutler Formation sandstone. Picture from Glen Canyon National Recreation Area, Utah.

The principle of original horizontality states that the deposition of sediments occurs as essentially horizontal beds. Observation of modern marine and non-marine sediments in a wide variety of environments supports this generalization (although cross-bedding is inclined, the overall orientation of cross-bedded units is horizontal).

The principle of superposition states that a sedimentary rock layer in a tectonically undisturbed sequence is younger than the one beneath it and older than the one above it. Logically a younger layer cannot slip beneath a layer previously deposited. This principle allows sedimentary layers to be viewed as a form of vertical time line, a partial or complete record of the time elapsed from deposition of the lowest layer to deposition of the highest bed.

The principle of faunal succession is based on the appearance of fossils in sedimentary rocks. As organisms exist at the same time period throughout the world, their presence or (sometimes) absence may be used to provide a relative age of the formations in which they are found. Based on principles laid out by William Smith almost a hundred years before the publication of Charles Darwin's theory of evolution, the principles of succession were developed independently of evolutionary thought. The principle becomes quite complex, however, given the uncertainties of fossilization, the localization of fossil types due to lateral changes in habitat (facies change in sedimentary strata), and that not all fossils may be found globally at the same time.

Absolute dating

Geologists also use methods to determine the absolute age of rock samples and geological events. These dates are useful on their own and may also be used in conjunction with relative dating methods or to calibrate relative methods.

At the beginning of the 20th century, important advancement in geological science was facilitated by the ability to obtain accurate absolute dates to geologic events using radioactive isotopes and other methods. This changed the understanding of geologic time. Previously, geologists could only use fossils and stratigraphic correlation to date sections of rock relative to one another. With isotopic dates it became possible to assign absolute ages to rock units, and these absolute dates could be applied to fossil sequences in which there was datable material, converting the old relative ages into new absolute ages.

For many geologic applications, isotope ratios of radioactive elements are measured in minerals that give the amount of time that has passed since a rock passed through its particular closure temperature, the point at which different radiometric isotopes stop diffusing into and out of the crystal lattice. These are used in geochronologic and thermochronologic studies. Common methods include uranium-lead dating, potassium-argon dating, argon-argon dating and uranium-thorium dating. These methods are used for a variety of applications. Dating of lava and volcanic ash layers found within a stratigraphic sequence can provide absolute age data for sedimentary rock units which do not contain radioactive isotopes and calibrate relative dating techniques. These methods can also be used to determine ages of pluton emplacement. Thermochemical techniques can be used to determine temperature profiles within the crust, the uplift of mountain ranges, and paleotopography.

Fractionation of the lanthanide series elements is used to compute ages since rocks were removed from the mantle.

Other methods are used for more recent events. Optically stimulated luminescence and cosmogenic radionucleide dating are used to date surfaces and/or erosion rates. Dendrochronology can also be used for the dating of landscapes. Radiocarbon dating is used for geologically young materials containing organic carbon.

GEOLOGICAL DEVELOPMENT OF AN AREA

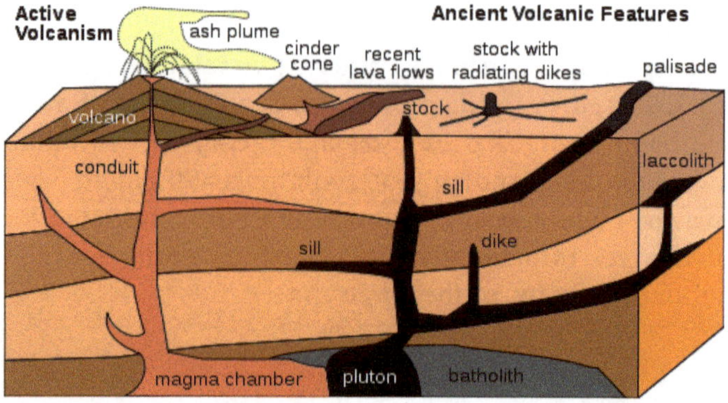

An originally horizontal sequence of sedimentary rocks (in shades of tan) are affected by igneous activity. Deep below the surface are a magma chamber and large associated igneous bodies. The magma chamber feeds the volcano, and sends offshoots of magma that will later crystallize into dikes and sills. Magma also advances upwards to form intrusive igneous bodies. The diagram illustrates both a cinder cone volcano, which releases ash, and a composite volcano, which releases both lava and ash.

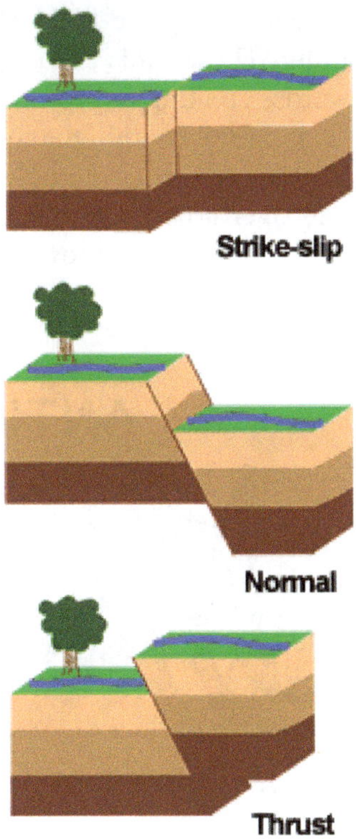

An illustration of the three types of faults. Strike-slip faults occur when rock units slide past one another, normal faults occur when rocks are undergoing horizontal extension, and thrust faults occur when rocks are undergoing horizontal shortening.

The geology of an area changes through time as rock units are deposited and inserted and deformational processes change their shapes and locations.

Rock units are first emplaced either by deposition onto the surface or intrusion into the overlying rock. Deposition can occur when sediments settle onto the surface of the Earth and later lithify into sedimentary rock, or when as volcanic material such as volcanic ash or lava flows blanket the surface. Igneous intrusions such as batholiths, laccoliths, dikes, and sills, push upwards into the overlying rock, and crystallize as they intrude.

After the initial sequence of rocks has been deposited, the rock units can be deformed and/or metamorphosed. Deformation typically occurs as a result of horizontal shortening, horizontal extension, or side-to-side (strike-slip) motion. These structural regimes broadly relate to convergent boundaries, divergent boundaries, and transform boundaries, respectively, between tectonic plates.

When rock units are placed under horizontal compression, they shorten and become thicker. Because rock units, other than muds, do not significantly change in volume, this is accomplished in two primary ways: through faulting and folding. In the shallow crust, where brittle deformation can occur, thrust faults form, which cause deeper rock to move on top of shallower rock. Because deeper rock is often older, as noted by the principle of superposition, this can result in older rocks moving on top of younger ones. Movement along faults can result in folding, either because the faults are not planar or because rock layers are dragged along, forming drag folds as slip occurs along the fault. Deeper in the Earth, rocks behave plastically, and fold instead of faulting. These folds can either be those where the material in the center of the fold buckles upwards, creating antiforms, or where it buckles downwards, creating synforms. If the tops of the rock units within the folds remain pointing upwards, they are called anticlines and synclines, respectively. If some of the units in the fold are facing downward, the structure is called an overturned anticline or syncline, and if all of the rock units are overturned or the correct up-direction is unknown, they are simply called by the most general terms, antiforms and synforms.

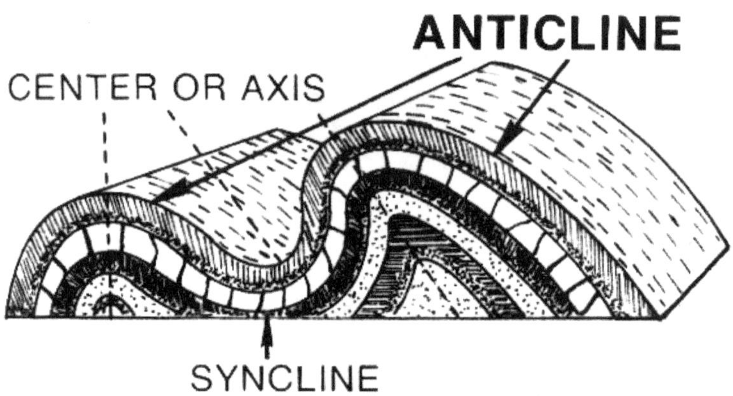

A diagram of folds, indicating an anticline and a syncline.

Even higher pressures and temperatures during horizontal shortening can cause both folding and metamorphism of the rocks. This metamorphism causes changes in the mineral composition of the rocks; creates a foliation, or planar surface, that is related to mineral growth under stress. This can remove signs of the original textures of the rocks, such as bedding in sedimentary rocks, flow features of lavas, and crystal patterns in crystalline rocks.

Extension causes the rock units as a whole to become longer and thinner. This is primarily accomplished through normal faulting and through the ductile stretching and thinning. Normal faults drop rock units that are higher below those that are lower. This typically results in younger units being placed below older units. Stretching of units can result in their thinning; in fact, there is a location within the Maria Fold and Thrust Belt in which the entire sedimentary sequence of the Grand Canyon can be seen over a length of less than a meter. Rocks at the depth to be ductilely stretched are often also metamorphosed. These stretched rocks can also pinch into lenses, known as boudins, after the French word for sausage, because of their visual similarity.

Where rock units slide past one another, strike-slip faults develop in shallow regions, and become shear zones at deeper depths where the rocks deform ductilely.

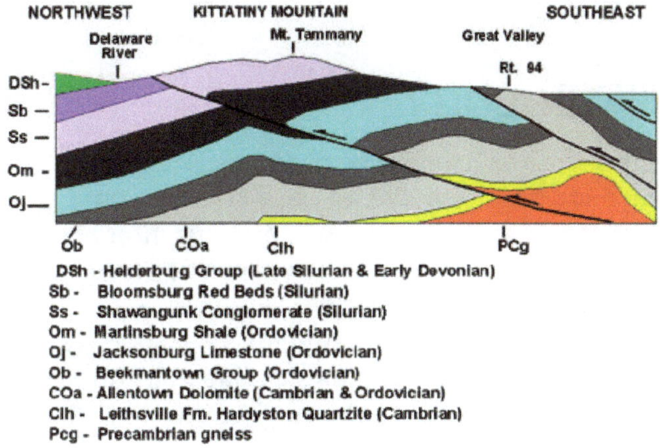

Geologic cross section of Kittatinny Mountain. This cross section shows metamorphic rocks, overlain by younger sediments deposited after the metamorphic event. These rock units were later folded and faulted during the uplift of the mountain.

The addition of new rock units, both depositionally and intrusively, often occurs during deformation. Faulting and other deformational processes result in the creation of topographic gradients, causing material on the rock unit that is increasing in elevation to be eroded by hillslopes and channels. These sediments are deposited on the rock unit that is going down. Continual motion along the fault maintains the topographic gradient in spite of the movement of sediment, and continues to create accommodation space for the material to deposit. Deformational events are often also associated with volcanism and igneous activity. Volcanic ashes and lavas accumulate on the surface, and igneous intrusions enter from below. Dikes, long, planar igneous intrusions, enter along cracks, and therefore often form in large numbers in areas that are being actively deformed. This can result in the emplacement of dike swarms, such as those that are observable across the Canadian shield, or rings of dikes around the lava tube of a volcano.

All of these processes do not necessarily occur in a single environment, and do not necessarily occur in a single order. The Hawaiian Islands, for example, consist almost entirely of layered basaltic lava flows. The sedimentary sequences of the mid-continental United States and the Grand Canyon in the southwestern United States contain almost-undeformed stacks of sedimentary rocks that have remained in place since Cambrian time. Other areas are much more geologically complex. In the southwestern United States, sedimentary, volcanic, and intrusive rocks have been metamorphosed, faulted, foliated, and folded. Even older rocks, such as the Acasta gneiss of the Slave craton in northwestern Canada, the oldest known rock in the world have been metamorphosed to the point where their origin is undiscernable without laboratory analysis. In addition, these processes can occur in stages. In many places, the Grand Canyon in the southwestern United States being a very visible example, the lower rock units were metamorphosed and deformed, and then deformation ended and the upper, undeformed units were deposited. Although any amount of rock emplacement and rock deformation can occur, and they can occur any number of times, these concepts provide a guide to understanding the geological history of an area.

METHODS OF GEOLOGY

Geologists use a number of field, laboratory, and numerical modeling methods to decipher Earth history and to understand the processes that occur on and inside the Earth. In typical geological investigations, geologists use primary information related to petrology (the study of rocks), stratigraphy (the study of sedimentary layers), and structural geology (the study of positions of rock units and their deformation). In many cases, geologists also study modern soils, rivers, landscapes, and glaciers; investigate past and current life and biogeochemical pathways, and use geophysical methods to investigate the subsurface. Sub-specialities of geology may distinguish **endogenous** and **exogenous** geology.

Field Methods

A standard Brunton Pocket Transit, commonly used by geologists for mapping and surveying.

A typical USGS field mapping camp in the 1950s

Today, handheld computers with GPS and geographic information systems software are often used in geological field work (digital geologic mapping).

Geological field work varies depending on the task at hand. Typical fieldwork could consist of:

- Geological mapping
    - Structural mapping: identifying the locations of major rock units and the faults and folds that led to their placement there.
    - Stratigraphic mapping: pinpointing the locations of sedimentary facies (lithofacies and biofacies) or the mapping of isopachs of equal thickness of sedimentary rock

    - Surficial mapping: recording the locations of soils and surficial deposits
- Surveying of topographic features
    - compilation of topographic maps
    - Work to understand change across landscapes, including:
        - Patterns of erosion and deposition
        - River-channel change through migration and avulsion
        - Hillslope processes
- Subsurface mapping through geophysical methods
    - These methods include:
        - Shallow seismic surveys
        - Ground-penetrating radar
        - Aeromagnetic surveys
        - Electrical resistivity tomography
    - They aid in:
        - Hydrocarbon exploration
        - Finding groundwater
        - Locating buried archaeological artifacts
- High-resolution stratigraphy
    - Measuring and describing stratigraphic sections on the surface
    - Well drilling and logging
- Biogeochemistry and geomicrobiology
    - Collecting samples to:
        - determine biochemical pathways
        - identify new species of organisms
        - identify new chemical compounds
    - and to use these discoveries to:

    - understand early life on Earth and how it functioned and metabolized
    - find important compounds for use in pharmaceuticals
- Paleontology: excavation of fossil material
    - For research into past life and evolution
    - For museums and education
- Collection of samples for geochronology and thermochronology
- Glaciology: measurement of characteristics of glaciers and their motion

Petrology

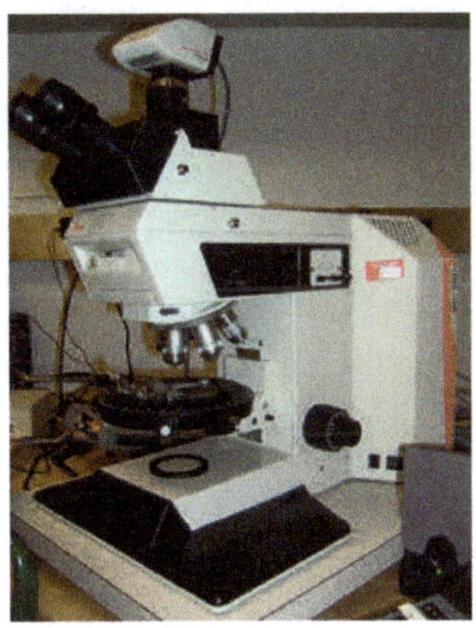

A petrographic microscope - an optical microscope fitted with cross-polarizing lenses, a conoscopic lens, and compensators (plates of anisotropic materials; gypsum plates and quartz wedges are common), for crystallographic analysis.

In addition to identifying rocks in the field (lithology), petrologists identify rock samples in the laboratory. Two of the primary methods for identifying rocks in the laboratory are through optical microscopy and by using an electron microprobe. In an optical mineralogy analysis, petrologists analyze thin sections of rock samples using a petrographic microscope, where the minerals can be identified through their different properties in plane-polarized and cross-polarized light, including their birefringence, pleochroism, twinning, and interference properties with a conoscopic lens. In the electron microprobe, individual locations are analyzed for their exact chemical compositions and variation in composition within individual crystals. Stable and radioactive isotope studies provide insight into the geochemical evolution of rock units.

Petrologists can also use fluid inclusion data and perform high temperature and pressure physical experiments to understand the temperatures and pressures at which different mineral phases appear, and how they change through igneous and metamorphic processes. This research can be extrapolated to the field to understand metamorphic processes and the conditions of crystallization of igneous rocks. This work can also help to explain processes that occur within the Earth, such as subduction and magma chamber evolution.

Structural Geology

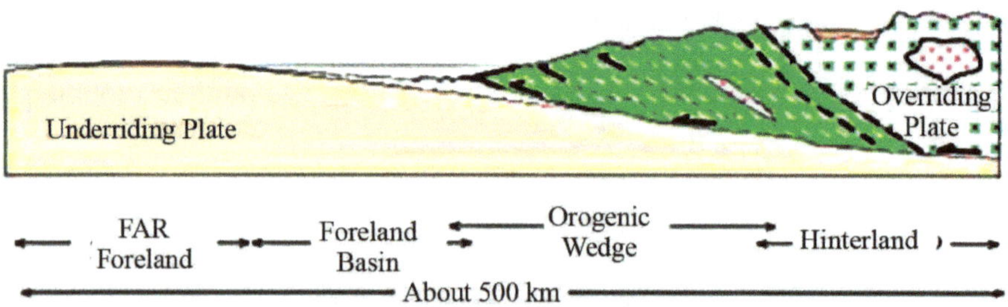

A diagram of an orogenic wedge. The wedge grows through faulting in the interior and along the main basal fault, called the décollement. It builds its shape into a critical taper, in which the angles within the wedge remain the same as failures inside the material balance failures along the décollement. It is analogous to a bulldozer pushing a pile of dirt, where the bulldozer is the overriding plate.

Structural geologists use microscopic analysis of oriented thin sections of geologic samples to observe the fabric within the rocks which gives information about strain within the crystalline structure of the rocks. They also plot and combine measurements of geological structures in order to better understand the orientations of faults and folds in order to reconstruct the history of rock deformation in the area. In addition, they perform analog and numerical experiments of rock deformation in large and small settings.

The analysis of structures is often accomplished by plotting the orientations of various features onto stereonets. A stereonet is a stereographic projection of a sphere onto a plane, in which planes are projected as lines and lines are projected as points. These can be used to find the locations of fold axes, relationships between faults, and relationships between other geologic structures.

Among the most well-known experiments in structural geology are those involving orogenic wedges, which are zones in which mountains are built along convergent tectonic plate boundaries. In the analog versions of these experiments, horizontal layers of sand are pulled along a lower surface into a back stop, which results in realistic-looking patterns of faulting and the growth of a critically tapered (all angles remain the same) orogenic wedge. Numerical models work in the same way as these analog models, though they are often more sophisticated and can include patterns of erosion and uplift in the mountain belt. This helps to show the relationship between erosion and the shape of a mountain range. These studies can also give useful information about pathways for metamorphism through pressure, temperature, space, and time.

Stratigraphy

In the laboratory, stratigraphers analyze samples of stratigraphic sections that can be returned from the field, such as those from drill cores. Stratigraphers also analyze data from geophysical surveys that show the locations of stratigraphic units in the subsurface. Geophysical data and well logs can be combined to produce a better view of the subsurface, and stratigraphers often use computer programs to do this in three dimensions. Stratigraphers can then use these data to reconstruct ancient processes occurring on the surface of the Earth, interpret past environments, and locate areas for water, coal, and hydrocarbon extraction.

In the laboratory, biostratigraphers analyze rock samples from outcrop and drill cores for the fossils found in them. These fossils help scientists to date the core and to understand the depositional environment in which the rock units formed. Geochronologists precisely date rocks within the stratigraphic section in order to provide better absolute bounds on the timing and rates of deposition. Magnetic stratigraphers look for signs of magnetic reversals in igneous rock units within the drill cores. Other scientists perform stable-isotope studies on the rocks to gain information about past climate.

PLANETARY GEOLOGY

Surface of Mars as photographed by the Viking 2 lander December 9, 1977.

With the advent of space exploration in the twentieth century, geologists have begun to look at other planetary bodies in the same ways that have been developed to study the Earth. This new field of study is called planetary geology (sometimes known as astrogeology) and relies on known geologic principles to study other bodies of the solar system. Although the Greek-language-origin prefix *geo* refers to Earth, "geology" is often used in conjunction with the names of other planetary bodies when describing their composition and internal processes: examples are "the geology of Mars and Lunar geology. Specialised terms such as *selenology* (studies of the Moon), *areology* (of Mars), etc., are also in use.

Although planetary geologists are interested in studying all aspects of other planets, a significant focus is to search for evidence of past or present life on other worlds. This has led to many missions whose primary or ancillary purpose is to examine planetary bodies for evidence of life. One of these is the Phoenix lander, which analyzed Martian polar soil for water, chemical, and mineralogical constituents related to biological processes.

APPLIED GEOLOGY

Economic Geology

Economic geology is an important branch of geology which deals with different aspects of economic minerals being used by humankind to fulfill its various needs. The economic minerals are those which can be extracted profitably. Economic geologists help locate and manage the Earth's natural resources, such as petroleum and coal, as well as mineral resources, which include metals such as iron, copper, and uranium.

Mining geology

Mining geology consists of the extractions of mineral resources from the Earth. Some resources of economic interests include gemstones, metals, and many minerals such as asbestos, perlite, mica, phosphates, zeolites, clay, pumice, quartz, and silica, as well as elements such as sulfur, chlorine, and helium.

Petroleum Geology

Mud log in process, a common way to study the lithology when drilling oil wells.

Petroleum geologists study the locations of the subsurface of the Earth which can contain extractable hydrocarbons, especially petroleum and natural gas. Because many of these reservoirs are found in sedimentary basins, they study the formation of these basins, as well as their sedimentary and tectonic evolution and the present-day positions of the rock units.

Engineering Geology

Engineering geology is the application of the geologic principles to engineering practice for the purpose of assuring that the geologic factors affecting the location, design, construction, operation, and maintenance of engineering works are properly addressed.

In the field of civil engineering, geological principles and analyses are used in order to ascertain the mechanical principles of the material on which structures are built. This allows tunnels to be built without collapsing, bridges and skyscrapers to be built with sturdy foundations, and buildings to be built that will not settle in clay and mud.

Hydrology and Environmental Issues

Geology and geologic principles can be applied to various environmental problems such as stream restoration, the restoration of brownfields, and the understanding of the interaction between natural habitat and the geologic environment. Groundwater hydrology, or hydrogeology, is used to locate groundwater, which can often provide a ready supply of uncontaminated water and is especially important in arid regions, and to monitor the spread of contaminants in groundwater wells.

Geologists also obtain data through stratigraphy, boreholes, core samples, and ice cores. Ice cores and sediment cores are used to for paleoclimate reconstructions, which tell geologists about past and present temperature, precipitation, and sea level across the globe. These datasets are our primary source of information on global climate change outside of instrumental data.

Natural Hazards

Geologists and geophysicists study natural hazards in order to enact safe building codes and warning systems that are used to prevent loss of property and life. Examples of important natural hazards that are pertinent to geology (as opposed those that are mainly or only pertinent to meteorology) are:

Rockfall in the Grand Canyon

- Avalanches
- Earthquakes
- Floods
- Landslides and debris flows
- River channel migration and avulsion
- Liquefaction
- Sinkholes
- Subsidence
- Tsunamis
- Volcanoes

HISTORY OF GEOLOGY

The study of the physical material of the Earth dates back at least to ancient Greece when Theophrastus (372–287 BCE) wrote the work *Peri Lithon* (*On Stones*). During the Roman period, Pliny the Elder wrote in detail of the many minerals and metals then in practical use – even correctly noting the origin of amber.

Some modern scholars, such as Fielding H. Garrison, are of the opinion that the origin of the science of geology can be traced to Persia after the Muslim conquests had come to an end. Abu al-Rayhan al-Biruni (973–1048 CE) was one of the earliest Persian geologists, whose works included the earliest writings on the geology of India, hypothesizing that the Indian subcontinent was once a sea.

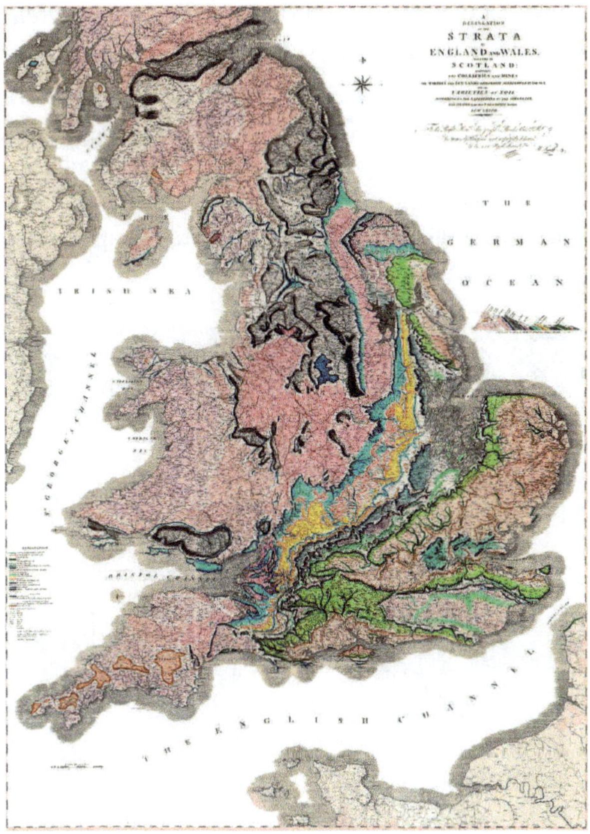

William Smith's geologic map of England, Wales, and southern Scotland. Completed in 1815, it was the second national-scale geologic map, and by far the most accurate of its time.

Drawing from Greek and Indian scientific literature that were not destroyed by the Muslim conquests, the Persian scholar Ibn Sina (Avicenna, 981–1037) proposed detailed explanations for the formation of mountains, the origin of earthquakes, and other topics central to modern geology, which provided an essential foundation for the later development of the science. In China, the polymath Shen Kuo (1031–1095) formulated a hypothesis for the process of land formation: based on his observation of fossil animal shells in a geological stratum in a mountain hundreds of miles from the ocean, he inferred that the land was formed by erosion of the mountains and by deposition of silt.

Nicolas Steno (1638–1686) is credited with the law of superposition, the principle of original horizontality, and the principle of lateral continuity: three defining principles of stratigraphy.

The word *geology* was first used by Ulisse Aldrovandi in 1603, then by Jean-André Deluc in 1778 and introduced as a fixed term by Horace-Bénédict de Saussure in 1779. The word is derived from the Greek γ, *gê*, meaning "earth" and λόγος, *logos*, meaning "speech". But according to another source, the word "geology" comes from a Norwegian, Mikkel Pedersøn Escholt (1600–1699), who was a priest and scholar. Escholt first used the definition in his book titled, *Geologica Norvegica* (1657).

William Smith (1769–1839) drew some of the first geological maps and began the process of ordering rock strata (layers) by examining the fossils contained in them.

James Hutton is often viewed as the first modern geologist. In 1785 he presented a paper entitled *Theory of the Earth* to the Royal Society of Edinburgh. In his paper, he explained his theory that the Earth must be much older than had previously been supposed in order to allow enough time for mountains to be eroded and for sediments to form new rocks at the bottom of the sea, which in turn were raised up to become dry land. Hutton published a two-volume version of his ideas in 1795 (Vol. 1, Vol. 2).

Scotsman James Hutton, father of modern geology

Followers of Hutton were known as *Plutonists* because they believed that some rocks were formed by *vulcanism*, which is the deposition of lava from volcanoes, as opposed to the *Neptunists*, led by Abraham Werner, who believed that all rocks had settled out of a large ocean whose level gradually dropped over time.

The first geological map of the U.S. was produced in 1809 by William Maclure. In 1807, Maclure commenced the self-imposed task of making a geological survey of the United States. Almost every state in the Union was traversed and mapped by him, the Allegheny Mountains being crossed and recrossed some 50 times. The results of his unaided labours were submitted to the American Philosophical Society in a memoir entitled *Observations on the Geology of the United States explanatory of a Geological Map*, and published in the *Society's Transactions*, together with the nation's first geological map.

This antedates William Smith's geological map of England by six years, although it was constructed using a different classification of rocks.

Sir Charles Lyell first published his famous book, *Principles of Geology*, in 1830. This book, which influenced the thought of Charles Darwin, successfully promoted the doctrine of uniformitarianism. This theory states that slow geological processes have occurred throughout the Earth's history and are still occurring today. In contrast, catastrophism is the theory that Earth's features formed in single, catastrophic events and remained unchanged thereafter. Though Hutton believed in uniformitarianism, the idea was not widely accepted at the time.

Much of 19th-century geology revolved around the question of the Earth's exact age. Estimates varied from a few hundred thousand to billions of years. By the early 20th century, radiometric dating allowed the Earth's age to be estimated at two billion years. The awareness of this vast amount of time opened the door to new theories about the processes that shaped the planet.

Some of the most significant advances in 20th-century geology have been the development of the theory of plate tectonics in the 1960s and the refinement of estimates of the planet's age. Plate tectonics theory arose from two separate geological observations: seafloor spreading and continental drift. The theory revolutionized the Earth sciences. Today the Earth is known to be approximately 4.5 billion years old.

REFERENCES

1. "geology". Online Etymology Dictionary.
2. γ. Liddell, Henry George; Scott, Robert; A Greek–English Lexicon at the Perseus Project
3. New Hampshire Geological Survey, Geologic maps, section Surficial Geologic Maps
4. Hess, H. H. (November 1, 1962) "History Of Ocean Basins", pp. 599–620 in Petrologic studies: a volume in honor of A. F. Buddington. A. E. J. Engel, Harold L. James, and B. F. Leonard (eds.) [New York?]: Geological Society of America.
5. Kious, Jacquelyne; Tilling, Robert I. (February 1996). Developing the Theory. This Dynamic Earth: The Story of Plate Tectonics. Kiger, Martha, Russel, Jane (Online ed.). Reston, Virginia, USA: United States Geological Survey. ISBN 0-16-048220-8. Retrieved 13 March 2009.
6. Kious, Jacquelyne; Tilling, Robert I. (February 1996). Understanding Plate Motions. This Dynamic Earth: The Story of Plate Tectonics. Kiger, Martha, Russel, Jane (Online ed.). Reston, Virginia, USA: United States Geological Survey. ISBN 0-16-048220-8. Retrieved 13 March 2009.
7. Wegener, A. (1999). Origin of continents and oceans. Courier Corporation. ISBN 0-486-61708-4.
8. International Commission on Stratigraphy Archived September 20, 2005, at the Wayback Machine.. stratigraphy.org

9. Amelin, Y. (2002). Lead Isotopic Ages of Chondrules and Calcium-Aluminum-Rich Inclusions. Science. 297 *(5587): 1678–83. Bibcode*:2002Sci...297.1678A. doi:10.1126/science.1073950. PMID 12215641.

10. Patterson, C. (1956). Age of Meteorites and the Earth. Geochimica et Cosmochimica Acta. 10: 230–237. Bibcode:1956GeCoA..10..230P. doi:10.1016/0016-7037(56)90036-9.

11. Dalrymple, G. Brent (1994). The age of the earth. Stanford, California: Stanford Univ. Press. ISBN 0-8047-2331-1.

12. Reijer Hooykaas, Natural Law and Divine Miracle: The Principle of Uniformity in Geology, Biology, and Theology, Leiden: EJ Brill, 1963.

13. Levin, Harold L. (2010). The earth through time (9th ed.). Hoboken, N.J.: J. Wiley. p. 18. ISBN 978-0-470-38774-0.

14. Olsen, Paul E. (2001). Steno's Principles of Stratigraphy. Dinosaurs and the History of Life. Columbia University. Retrieved 2009-03-14.

15. As recounted in Simon Winchester, The Map that Changed the World (New York: HarperCollins, 2001) pp. 59–91.

16. Tucker, R. D.; Bradley, D. C.; Ver Straeten, C. A.; Harris, A. G.; Ebert, J. R.; McCutcheon, S. R. (1998). New U–Pb zircon ages and the duration and division of Devonian time. Earth and Planetary Science Letters. 158 *(3–4): 175–186. Bibcode*:1998E&PSL.158..175T. doi:10.1016/S0012-821X(98)00050-8.

17. Rollinson, Hugh R. (1996). Using geochemical data evaluation, presentation, interpretation. Harlow: Longman. ISBN 978-0-582-06701-1.

18. Faure, Gunter (1998). Principles and applications of geochemistry: a comprehensive textbook for geology students. Upper Saddle River, NJ: Prentice-Hall. ISBN 978-0-02-336450-1.

19. Compare: Hansen, Jens Morten (2009). On the origin of natural history: Steno's modern, but forgotten philosophy of science. In Rosenberg, Gary D. The Revolution in Geology from the Renaissance to the Enlightenment. Geological Society of America Memoir. 203. Boulder, Colorado: Geological Society of America. p. 169. ISBN 9780813712031. Retrieved 2016-08-24. [...] the historic dichotomy between ‹hard rock' and ‹soft rock' geologists, i.e. scientists working mainly with endogenous and exogenous processes, respectively [...] endogenous forces mainly defining the developments below Earth's crust and the exogenous forces mainly defining the developments on top of and above Earth's crust.

20. Compton, Robert R. (1985). Geology in the field. New York: Wiley. ISBN 0-471-82902-1.

21. USGS Topographic Maps. United States Geological Survey. Retrieved 2009-04-11.

22. Burger, H. Robert; Sheehan, Anne F.; Jones, Craig H. (2006). Introduction to applied geophysics : exploring the shallow subsurface. New York: W.W. Norton. ISBN 0-393-92637-0.

23. Krumbein, Wolfgang E., ed. (1978). Environmental biogeochemistry and

geomicrobiology. Ann Arbor, Mich.: Ann Arbor Science Publ. ISBN 0-250-40218-1.

24. McDougall, Ian; Harrison, T. Mark (1999). Geochronology and thermochronology by the °Ar/©Ar method. New York: Oxford University Press. ISBN 0-19-510920-1.

25. Hubbard, Bryn; Glasser, Neil (2005). Field techniques in glaciology and glacial geomorphology. Chichester, England: J. Wiley. ISBN 0-470-84426-4.

26. Nesse, William D. (1991). Introduction to optical mineralogy. New York: Oxford University Press. ISBN 0-19-506024-5.

27. Morton, A. C. (1985). A new approach to provenance studies: electron microprobe analysis of detrital garnets from Middle Jurassic sandstones of the northern North Sea. Sedimentology. 32 *(4): 553–566. Bibcode*:1985Sedim..32..553M. doi:10.1111/j.1365-3091.1985.tb00470.x.

28. Zheng, Y; Fu, Bin; Gong, Bing; Li, Long (2003). Stable isotope geochemistry of ultrahigh pressure metamorphic rocks from the Dabie–Sulu orogen in China: implications for geodynamics and fluid regime. Earth-Science Reviews. 62: 105–161. Bibcode:2003ESRv...62..105Z. doi:10.1016/S0012-8252(02)00133-2.

29. Condomines, M; Tanguy, J; Michaud, V (1995). Magma dynamics at Mt Etna: Constraints from U-Th-Ra-Pb radioactive disequilibria and Sr isotopes in historical lavas. Earth and Planetary Science Letters. 132: 25–41. Bibcode:1995E&PSL.132...25C. doi:10.1016/0012-821X(95)00052-E.

30. Shepherd, T.J.; Rankin, A.H.; Alderton, D.H.M. (1985). A practical guide to fluid inclusion studies. Glasgow: Blackie. ISBN 0-412-00601-4.

31. Sack, Richard O.; Walker, David; Carmichael, Ian S. E. (1987). Experimental petrology of alkalic lavas: constraints on cotectics of multiple saturation in natural basic liquids. Contributions to Mineralogy and Petrology. 96: 1–23. Bibcode:1987CoMP...96....1S. doi:10.1007/BF00375521.

32. McBirney, Alexander R. (2007). Igneous petrology. Boston: Jones and Bartlett Publishers. ISBN 978-0-7637-3448-0.

33. Spear, Frank S. (1995). Metamorphic phase equilibria and pressure-temperature-time paths. Washington, DC: Mineralogical Soc. of America. ISBN 978-0-939950-34-8.

34. Dahlen, F A (1990). Critical Taper Model of Fold-And-Thrust Belts and Accretionary Wedges. Annual Review of Earth and Planetary Sciences. 18: 55–99. Bibcode:1990AREPS..18...55D. doi:10.1146/annurev.ea.18.050190.000415.

35. Gutscher, M; Kukowski, Nina; Malavieille, Jacques; Lallemand, Serge (1998). Material transfer in accretionary wedges from analysis of a systematic series of analog experiments. Journal of Structural Geology. 20 *(4): 407–416. Bibcode*:1998JSG....20..407G. doi:10.1016/S0191-8141(97)00096-5.

36. Koons, P O (1995). Modeling the Topographic Evolution of Collisional Belts. Annual Review of Earth and Planetary Sciences. 23: 375–408. Bibcode:1995AREPS..23..375K. doi:10.1146/annurev.ea.23.050195.002111.

37. Dahlen, F. A.; Suppe, J.; Davis, D. (1984). Mechanics of Fold-and-Thrust Belts and Accretionary Wedges: Cohesive Coulomb Theory. J. Geophys. Res. 89 *(B12): 10087–10101. Bibcode*:1984JGR....8910087D. doi:10.1029/JB089iB12p10087.
38. Hodell, David A.; Benson, Richard H.; Kent, Dennis V.; Boersma, Anne; Rakic-El Bied, Kruna (1994). Magnetostratigraphic, Biostratigraphic, and Stable Isotope Stratigraphy of an Upper Miocene Drill Core from the Salé Briqueterie (Northwestern Morocco): A High-Resolution Chronology for the Messinian Stage. Paleoceanography. 9 *(6): 835–855. Bibcode*:1994PalOc...9..835H. doi:10.1029/94PA01838.
39. Bally, A.W., ed. (1987). Atlas of seismic stratigraphy. Tulsa, Okla., U.S.A.: American Association of Petroleum Geologists. ISBN 0-89181-033-1.
40. Fernández, O.; Muñoz, J. A.; Arbués, P.; Falivene, O.; Marzo, M. (2004). Three-dimensional reconstruction of geological surfaces: An example of growth strata and turbidite systems from the Ainsa basin (Pyrenees, Spain). AAPG Bulletin. 88 *(8): 1049–1068.* doi:10.1306/02260403062.
41. Poulsen, Chris J.; Flemings, Peter B.; Robinson, Ruth A. J.; Metzger, John M. (1998). Three-dimensional stratigraphic evolution of the Miocene Baltimore Canyon region: Implications for eustatic interpretations and the systems tract model. Geological Society of America Bulletin. 110 *(9): 1105–1122. Bibcode*:1998GSAB..110.1105P. doi:10.1130/0016-7606(1998)110<1105:TDSEOT>2.3.CO;2.
42. Toscano, M; Lundberg, Joyce (1999). Submerged Late Pleistocene reefs on the tectonically-stable S.E. Florida margin: high-precision geochronology, stratigraphy, resolution of Substage 5a sea-level elevation, and orbital forcing. Quaternary Science Reviews. 18 *(6): 753–767. Bibcode*:1999QSRv...18..753T. doi:10.1016/S0277-3791(98)00077-8.
43. Selley, Richard C. (1998). Elements of petroleum geology. San Diego: Academic Press. ISBN 0-12-636370-6.
44. Das, Braja M. (2006). Principles of geotechnical engineering. England: Thomson Learning. ISBN 0-534-55144-0.
45. Hamilton, Pixie A.; Helsel, Dennis R. (1995). Effects of Agriculture on Ground-Water Quality in Five Regions of the United States. Ground Water. 33 *(2): 217–226. doi:10.1111/j.1745-6584.1995.tb00276.x.*
46. Seckler, David; Barker, Randolph; Amarasinghe, Upali (1999). Water Scarcity in the Twenty-first Century. International Journal of Water Resources Development. 15: 29–42. doi:10.1080/07900629948916.
47. Welch, Alan H.; Lico, Michael S.; Hughes, Jennifer L. (1988). Arsenic in Ground Water of the Western United States. Ground Water. 26 *(3): 333–347. doi:10.1111/j.1745-6584.1988.tb00397.x.*
48. Barnola, J. M.; Raynaud, D.; Korotkevich, Y. S.; Lorius, C. (1987). Vostok ice core provides 160,000-year record of atmospheric CO2. Nature. 329 *(6138): 408–414. Bibcode*:1987Natur.329..408B. doi:10.1038/329408a0.
49. Colman, S.M.; Jones, G.A.; Forester, R.M.; Foster, D.S. (1990). Holocene

paleoclimatic evidence and sedimentation rates from a core in southwestern Lake Michigan. Journal of Paleolimnology. 4 *(3)*. doi:10.1007/BF00239699.

50. Jones, P. D.; Mann, M. E. (6 May 2004). Climate over past millennia (PDF). Reviews of Geophysics. 42 *(2)*. *Bibcode*:2004RvGeo..42.2002J. doi:10.1029/2003RG000143.

51. USGS Natural Hazards Gateway. usgs.gov

52. Winchester, Simon (2002). The map that changed the world: William Smith and the birth of modern geology. New York, NY: Perennial. ISBN 0-06-093180-9.

53. "The Saracens themselves were the originators not only of algebra, chemistry, and geology, but of many of the so-called improvements or refinements of civilization, such as street lamps, window-panes, fireworks, stringed instruments, cultivated fruits, perfumes, spices, etc." (Fielding H. Garrison, An introduction to the history of medicine, W.B. Saunders, 1921, p. 116)

54. Asimov, M. S.; Bosworth, Clifford Edmund (eds.). The Age of Achievement: A.D. 750 to the End of the Fifteenth Century : The Achievements. History of civilizations of Central Asia. pp. 211–214. ISBN 978-92-3-102719-2.

55. Toulmin, S. and Goodfield, J. (1965) The Ancestry of science: The Discovery of Time, Hutchinson & Co., London, p. 64

56. Al-Rawi, Munin M. (November 2002). The Contribution of Ibn Sina (Avicenna) to the development of Earth Sciences (PDF) (Report). Manchester, UK: Foundation for Science Technology and Civilisation. Publication 4039. Retrieved April 2012. Check date values in: |access-date= (help)

57. Needham, Joseph (1986). Science and Civilization in China: Volume 3, Mathematics and the Sciences of the Heavens and the Earth. Taipei: Caves Books, Ltd. pp. 603–604.

58. Vai, Gian Battista; Cavazza, William (2003). Four centuries of the word geology: Ulisse Aldrovandi 1603 in Bologna. Minerva. ISBN 978-88-7381-056-8.

59. Winchester, Simon (2001). The Map that Changed the World. HarperCollins Publishers. p. 25. ISBN 0-06-093180-9.

60. Kermit H., (2003) Niels Stensen, 1638–1686: the scientist who was beatified. Gracewing Publishing. p. 127.

61. James Hutton: The Founder of Modern Geology American Museum of Natural History

62. Maclure, William (1817). Observations on the Geology of the United States of America: With Some Remarks on the Effect Produced on the Nature and Fertility of Soils, by the Decomposition of the Different Classes of Rocks; and an Application to the Fertility of Every State in the Union, in Reference to the Accompanying Geological Map ... author.

63. Observations on the Geology of the United States (Maclure). David Rumsey map collection

64. Greene, J.C.; Burke, J.G. (1978). The Science of Minerals in the Age of Jefferson. Transactions of the American Philosophical Society, New Series. 68

(4): 1–113 (39). doi:10.2307/1006294.

65. Maclure's 1809 Geological Map. davidrumsey.com
66. Lyell, Charles (1991). Principles of geology. Chicago: University of Chicago Press. ISBN 978-0-226-49797-6.
67. England, Philip; Molnar, Peter; Richter, Frank (2007). John Perry's neglected critique of Kelvin's age for the Earth: A missed opportunity in geodynamics. GSA Today. 17: 4. doi:10.1130/GSAT01701A.1.
68. Dalrymple, G.B. (1991). The Age of the Earth. California: Stanford University Press. ISBN 0-8047-1569-6.

CHAPTER 2

History of Geology

Chapter Contents

- **Introduction**
- **Antiquity**
- **Modern Geology**
- **Stratigraphy**
- **Geological History of Earth**
- **Timeline of Geology**

Only a small part of rocks are stable on earth´s surface, as widespread erosion and alteration shows. Magmatic rocks, like granite, crumble and minerals composing these rocks, like feldspar and mica, react with water to become soft clay. Similar things happen to metamorphic rocks, eroding under atmospheric conditions. Often metamorphites, formed deep within earth, quickly start to decay already during uplift. Especially unstable are metamorphic rocks formed under very high pressure (25-30 kbar) and temperature (650-700°C), conditions as found in earth´s crust in a depth ranging from 56 to 186miles (90-300km). One locality where such Ultra-High-Pressure (UHP) rocks are partially preserved are the Western Alps between Italy and France, in particular the Dora-Maira Massif.

INTRODUCTION

The history of geology is concerned with the development of the natural science of geology. Geology is the scientific study of the origin, history, and structure of the Earth.

ANTIQUITY

A mosquito and a fly in this Baltic amber necklace are between 40 and 60 million years old

The slightly misshapen octahedral shape of this rough diamond crystal in matrix is typical of the mineral. Its lustrous faces also indicate that this crystal is from a primary deposit.

Some of the first geological thoughts were about the origin of the Earth. Ancient Greece developed some primary geological concepts concerning the origin of the Earth. Additionally, in the 4th century BC Aristotle made critical observations of the slow rate of geological change. He observed the composition of the land and formulated a theory where the Earth changes at a slow rate and that these changes cannot be observed during one person's lifetime. Aristotle developed one of the first evidentially based concepts connected to the geological realm regarding the rate at which the Earth physically changes.

However, it was his successor at the Lyceum, the philosopher Theophrastus, who made the greatest progress in antiquity in his work *On Stones*. He described many minerals and ores both from local mines such as those at Laurium near Athens, and further afield. He also quite naturally discussed types of marble and building materials like limestones, and attempted a primitive classification of the properties of minerals by their properties such as hardness.

Much later in the Roman period, Pliny the Elder produced a very extensive discussion of many more minerals and metals then widely used for practical ends. He was among the first to correctly identify the origin of amber as a fossilized resin from trees by the observation of insects trapped within some pieces. He also laid the basis of crystallography by recognising the octahedral habit of diamond.

Middle Ages

Abu al-Rayhan al-Biruni (AD 973-1048) was one of the earliest Muslim geologists, whose works included the earliest writings on the geology of India, hypothesizing that the Indian subcontinent was once a sea:

Ibn Sina (Avicenna, 981-1037), a Persian polymath, made significant contributions to geology and the natural sciences (which he called *Attabieyat*) along with other natural philosophers such as Ikhwan AI-Safa and many others. Ibn Sina wrote an encyclopaedic work entitled "*Kitab al-Shifa*" (the Book of Cure, Healing or Remedy from ignorance), in which Part 2, Section 5, contains his commentary on Aristotle's Mineralogy and Meteorology, in six chapters: Formation of mountains, The advantages of mountains in the formation of clouds; Sources of water; Origin of earthquakes; Formation of minerals; The diversity of earth's terrain.

In medieval China, one of the most intriguing naturalists was Shen Kuo (1031-1095), a polymath personality who dabbled in many fields of study in his age. In terms of geology, Shen Kuo is one of the first naturalists to have formulated a theory of geomorphology. This was based on his observations of sedimentary uplift, soil erosion, deposition of silt, and marine fossils found in the Taihang Mountains, located hundreds of miles from the Pacific Ocean. He also formulated a theory of gradual climate change, after his observation of ancient petrified bamboos found in a preserved state underground near Yanzhou (modern Yan'an), in the dry northern climate of Shaanxi province. He formulated a hypothesis for the process of land formation: based on his observation of fossil shells in a geological stratum in a mountain hundreds of miles from the ocean, he inferred that the land was formed by erosion of the mountains and by deposition of silt.

17TH CENTURY

It was not until the 17th century that geology made great strides in its development. At this time, geology became its own entity in the world of natural science. It was discovered by the Christian world that different translations of the Bible contained different versions of the biblical text. The one entity that remained consistent through all of the interpretations was that the Deluge had formed the world's geology and geography. To prove the Bible's authenticity, individuals felt the need to demonstrate with scientific evidence that the Great Flood had in fact occurred. With this enhanced desire for data came an increase in observations of the Earth's composition, which in turn led to the discovery of fossils. Although theories that resulted from the heightened interest in the Earth's composition were often manipulated to support the concept of the Deluge, a genuine outcome was a greater interest in the makeup of the Earth. Due to the strength of Christian beliefs during the 17th century, the theory of the origin of the Earth that was most widely accepted was *A New Theory of the Earth* published in 1696, by William Whiston. Whiston used Christian reasoning to "prove" that the Great Flood had occurred and that the flood had formed the rock strata of the Earth.

A portrait of Whiston with a diagram demonstrating his theories of cometary catastrophism best described in *A New Theory of the Earth*

During the 17th century the heated debate between religion and science over the Earth's origin further propelled interest in the Earth and brought about more systematic identification techniques of the Earth's strata. The Earth's strata can be defined as horizontal layers of rock having approximately the same composition throughout. An important pioneer in the science was Nicolas Steno. Steno was trained in the classical texts on science; however, by 1659 he seriously questioned accepted knowledge of the natural world. Importantly, he questioned the idea that fossils grew in the ground, as well as common explanations of rock formation. His investigations and his subsequent conclusions on these topics have led scholars to consider him one of the founders of modern stratigraphy and geology.

18TH CENTURY

From this increased interest in the nature of the Earth and its origin, came a heightened attention to minerals and other components of the Earth's crust. Moreover, the increasing economic importance of mining in Europe during the mid to late 18th century made the possession of accurate knowledge about ores and their natural distribution vital. Scholars began to study the makeup of the Earth in a systematic manner, with detailed comparisons and descriptions not only of the land itself, but of the semi-precious metals it contained, which had great commercial value. For example, in 1774 Abraham Gottlob Werner published the book *Von den äusserlichen Kennzeichen der Fossilien (On the External Characters of Minerals),* which brought him widespread recognition because he presented a detailed system for identifying specific minerals based on external characteristics. The more efficiently productive land for mining could be identified and the semi-precious metals could be found, the more money could be made. This drive for economic gain propelled geology into the limelight and made it a popular subject to pursue. With an increased number of people studying it, came more detailed observations and more information about the Earth.

Also during the eighteenth century, aspects of the history of the Earth—namely the divergences between the accepted religious concept and factual evidence—once again became a popular topic for discussion in society. In 1749 the French naturalist Georges-Louis Leclerc, Comte de Buffon published his *Histoire Naturelle,* in which he attacked the popular Biblical accounts given by Whiston and other ecclesiastical theorists of the history of Earth. From experimentation with cooling globes, he found that the age of the Earth was not only 4,000 or 5,500 years as inferred from the Bible, but rather 75,000 years. Another individual who described the history of the Earth with reference to neither God nor the Bible was the philosopher Immanuel Kant, who published his *Universal Natural History and Theory of the Heavens (Allgemeine Naturgeschichte und Theorie des Himmels)* in 1755. From the works of these respected men, as well as others, it became acceptable by the mid eighteenth century to question the age of the Earth. This questioning represented a turning point in the study of the Earth. It was now possible to study the history of the Earth from a scientific perspective without religious preconceptions.

With the application of scientific methods to the investigation of the Earth's history, the study of geology could become a distinct field of science. To begin with, the terminology and definition of what constituted geological study had to be worked out. The term geology was first used technically in publications by two Genevan naturalists, Jean-André Deluc and Horace-Bénédict de Saussure, though geology was not well received as a term until it was taken up in the very influential compendium, the *Encyclopédie*, published beginning in 1751 by Denis Diderot. Once the term was established to denote the study of the Earth and its history, geology slowly became more generally recognized as a distinct science that could be taught as a field of study at educational institutions. In 1741 the best-known institution in the field of natural history, the National Museum of Natural History in France, created the first teaching position designated specifically for geology.This was an important step in further promoting knowledge of geology as a science and in recognizing the value of widely disseminating such knowledge.

By the 1770s chemistry was starting to play a pivotal role in the theoretical foundation of geology and two opposite theories with committed followers emerged. These contrasting theories offered differing explanations of how the rock layers of the Earth's surface had formed. One suggested that a liquid inundation, perhaps like the biblical deluge, had created all geological strata. The theory extended chemical theories that had been developing since the seventeenth century and was promoted by Scotland's John Walker, Sweden's Johan Gottschalk Wallerius and Germany's Abraham Werner. Of these names, Werner's views become internationally influential around 1800. He argued that the Earth's layers, including basalt and granite, had formed as a precipitate from an ocean that covered the entire Earth. Werner's system was influential and those who accepted his theory were known as Diluvianists or Neptunists. The Neptunist thesis was the most popular during the late eighteenth century, especially for those who were chemically trained. However, another thesis slowly gained currency from the 1780s forward. Instead of water, some mid eighteenth-century naturalists such as Buffon had suggested that strata had been formed through heat (or fire). The thesis was modified and expanded by the Scottish naturalist James Hutton during the 1780s. He argued against the theory of Neptunism, proposing instead the theory of based on heat. Those who followed this thesis during the early nineteenth century referred to this view as Plutonism: the formation of the Earth through the gradual solidification of a molten mass at a slow rate by the same processes that had occurred throughout history and continued in the present day. This led him to the conclusion that the Earth was immeasurably old and could not possibly be explained within the limits of the chronology inferred from the Bible. Plutonists believed that volcanic processes were the chief agent in rock formation, not water from a Great Flood.

19TH CENTURY

Bust of William Smith, in the Oxford University Museum of Natural History.

In the early 19th century the mining industry and Industrial Revolution stimulated the rapid development of the stratigraphic column - "the sequence of rock formations arranged according to their order of formation in time." In England. the mining surveyor William Smith, starting in the 1790s, found empirically that fossils were a highly effective means of distinguishing between otherwise similar formations of the landscape as he travelled the country working on the canal system and produced the first geological map of Britain. At about the same time, the French comparative anatomist Georges Cuvier assisted by his colleague Alexandre Brogniart at the École des Mines de Paris realized that the relative ages of fossils could be determined from a geological standpoint; in terms of what layer of rock the fossils are located and the distance these layers of rock are from the surface of the Earth. Through the synthesis of their findings, Brogniart and Cuvier realized that different strata could be identified by fossil contents and thus each stratum could be assigned to a unique position in a sequence.

Engraving from William Smith's 1815 monograph on identifying strata by fossils

After the publication of Cuvier and Brongniart's book, "Description Geologiques des Environs de Paris" in 1811, which outlined the concept, stratigraphy became very popular amongst geologists; many hoped to apply this concept to all the rocks of the Earth. During this century various geologists further refined and completed the stratigraphic column. For instance, in 1833 while Adam Sedgwick was mapping rocks that he had established were from the Cambrian Period, Charles Lyell was elsewhere suggesting a subdivision of the Tertiary Period; whilst Roderick Murchison, mapping into Wales from a different direction, was assigning the upper parts of Sedgewick's *Cambrian* to the lower parts of his own Silurian Period. The stratigraphic column was significant because it supplied a method to assign a relative age of these rocks by slotting them into different positions in their stratigraphical sequence. This created a global approach to dating the age of the Earth and allowed for further correlations to be drawn from similarities found in the makeup of the Earth's crust in various countries.

In early nineteenth-century Britain, catastrophism was adapted with the aim of reconciling geological science with religious traditions of the biblical Great Flood. In the early 1820s English geologists including William Buckland and Adam Sedgwick interpreted "diluvial deposits as the outcome of Noah's flood, but by the end of the decade they revised their opinions in favour of local inundations. Charles Lyell challenged catastrophism with the publication in 1830 of the first volume of his book *Principles of Geology* which presented a variety of geological evidence from England, France, Italy and Spain to prove Hutton's ideas of gradualism correct. He argued that most geological change had been very gradual in human history. Lyell provided evidence for Uniformitarianism; a geological doctrine that processes occur at the same rates in the present as they did in the past and account for all of the Earth's geological features. Lyell's works were popular and widely read, the concept of Uniformitarianism had taken a strong hold in geological society.

During the same time that the stratigraphic column was being completed, imperialism drove several countries in the early to mid 19th century to explore distant lands to expand their empires. This gave naturalists the opportunity to collect data on these voyages. In 1831 Captain Robert FitzRoy, given charge of the coastal survey expedition of HMS *Beagle*, sought a suitable naturalist to examine the land and give geological advice. This fell to Charles Darwin, who had just completed his BA degree and had accompanied Sedgwick on a two-week Welsh mapping expedition after taking his Spring course on geology. Fitzroy gave Darwin Lyell's *Principles of Geology*, and Darwin became Lyell's first disciple, inventively theorising on uniformitarian principles about the geological processes he saw, and challenging some of Lyell's ideas. He speculated about the Earth expanding to explain uplift, then on the basis of the idea that ocean areas sank as land was uplifted, theorised that coral atolls grew from fringing coral reefs round sinking volcanic islands. This idea was confirmed when the *Beagle* surveyed the Cocos (Keeling) Islands, and in 1842 he published his theory on *The Structure and Distribution of Coral Reefs*. Darwin's discovery of giant fossils helped to establish his reputation as a geologist, and his theorising about the causes of their extinction led to his theory of evolution by natural selection published in *On the Origin of Species* in 1859.

Economic motivations for the practical use of geological data caused governments to support geological research. During the 19th century the governments of several countries including Canada, Australia, Great Britain and the United States funded geological surveying that would produce geological maps of vast areas of the countries. Geological surveying provides the location of useful minerals and such information could be used to benefit the country's mining industry. With the government funding of geological research, more individuals could study geology with better technology and techniques, leading to the expansion of the field of geology.

In the 19th century, scientific inquiry had estimated the Age of the Earth in terms of millions of years. By the early 20th Century radiogenic isotopes had been discovered and Radiometric Dating had been developed. In 1911 Arthur Holmes dated a sample from Ceylon at 1.6 billion years old using lead isotopes. In 1921 ae yearly meeting of the British Association for the Advancement of Science came to a rough consensus that the Age of the Earth was a few billion years old, and that radiometric dating was credible. Holmes published The Age of the Earth, an Introduction to Geological Ideas in 1927 in which he presented a range of 1.6 to 3.0 billion years. Subsequent dating has taken the Age of the Earth to around 4.55 billion years. Theories that did not comply with the scientific evidence that established the age of the Earth could no longer be accepted.

20TH CENTURY

In 1862, the physicist William Thomson, 1st Baron Kelvin, published calculations that fixed the age of Earth at between 20 million and 400 million years. He assumed that Earth had formed as a completely molten object, and determined the amount of time it would take for the near-surface to cool to its present temperature. With the discovery of radioactive decay the age of the Earth was pushed back even further. Arthur Holmes, was a pioneer of geochronology. In 1913 Holmes was on the staff of Imperial College, when he published his famous book *The Age of the Earth* in which he argued strongly in favour of the use of radioactive dating methods rather than methods based on geological sedimentation or cooling of the earth (many people still clung to Lord Kelvin's calculations of less than 100 million years). Holmes estimated the oldest Archean rocks to be 1,600 million years, but did not speculate about the Earth's age. By this time the discovery of isotopes had complicated the calculations and he spent the next years grappling with these. His promotion of the theory over the next decades earned him the nickname of Father of Modern Geochronology. By 1927 he had revised this figure to 3,000 million years and in the 1940s to 4,500±100 million years, based on measurements of the relative abundance of uranium isotopes established by Alfred O. C. Nier. The general method is now known as the Holmes-Houterman model after Fritz Houtermans who published in the same year, 1946.The established age of the Earth has been refined since then but has not significantly changed.

Alfred Wegener, around 1925

In 1912 Alfred Wegener proposed the theory of Continental Drift. This theory suggests that the shapes of continents and matching coastline geology between some continents indicates they were joined together in the past and formed a single landmass known as Pangaea; thereafter they separated and drifted like rafts over the ocean floor, currently reaching their present position. Additionally, the theory of continental drift offered a possible explanation as to the formation of mountains; Plate Tectonics built on the theory of continental drift.

Unfortunately, Wegener provided no convincing mechanism for this drift, and his ideas were not generally accepted during his lifetime. Arthur Homes accepted Wegener's theory and provided a mechanism: mantle convection, to cause the continents to move. However, it was not until after the Second World War that new evidence started to accumulate that supported continental drift. There followed a period of 20 extremely exciting years where the Theory of Continental Drift developed from being believed by a few to being the cornerstone of modern Geology. Beginning in 1947 research found new evidence about the ocean floor, and in 1960 Bruce C. Heezen published the concept of mid-ocean ridges.Soon after this, Robert S. Dietz and Harry H. Hess proposed that the oceanic crust forms as the seafloor spreads apart along mid-ocean ridges in seafloor spreading. This was seen as confirmation of mantle convection and so the major stumbling block to the theory was removed. Geophysical evidence suggested lateral motion of continents and that oceanic crust is younger than continental crust. This geophysical evidence also spurred the hypothesis of paleomagnetism, the record of the orientation of the Earth's magnetic field recorded in magnetic minerals. British geophysicist S. K. Runcorn suggested the concept of paleomagnetism from his finding that the continents had moved relative to the Earth's magnetic poles. Tuzo Wilson, who was a promoter of the sea floor spreading hypothesis and continental drift from the very beginning, added the concept of transform faults to the model, completing the classes of fault types necessary to make the mobility of the plates on the globe function. A symposium on continental drift was held at the Royal Society of London in 1965 must be regarded as the official start of the acceptance of plate tectonics by the scientific community.The abstracts from the symposium are issued as Blacket, Bullard, Runcorn;1965.In this symposium, Edward Bullard and co-workers showed with a computer calculation how the continents along both sides of the Atlantic would best fit to close the ocean, which became known as the famous "Bullard's Fit". By the late 1960s the weight of the evidence available saw Continental Drift as the generally accepted theory.

MODERN GEOLOGY

By applying sound stratigraphic principles to the distribution of craters on the Moon, it can be argued that almost overnight, Gene Shoemaker took the study of the Moon away from Lunar astronomers and gave it to Lunar geologists.

In recent years, geology has continued its tradition as the study of the character and origin of the Earth, its surface features and internal structure. What changed in the later 20th century is the perspective of geological study. Geology was now studied using a more integrative approach, considering the Earth in a broader context encompassing the atmosphere, biosphere and hydrosphere. Satellites located in space that take wide scope photographs of the Earth provide such a perspective. In 1972, The Landsat Program, a series of satellite missions jointly managed by NASA and the U.S. Geological Survey, began supplying satellite images that can be geologically analyzed. These images can be used to map major geological units, recognize and correlate rock types for vast regions and track the movements of Plate Tectonics. A few applications of this data include the ability to produce geologically detailed maps, locate sources of natural energy and predict possible natural disasters caused by plate shifts.

STRATIGRAPHY

Stratigraphy is a branch of geology which studies rock layers (strata) and layering (stratification). It is primarily used in the study of sedimentary and layered volcanic rocks. Stratigraphy has two related subfields: lithologic stratigraphy or lithostratigraphy, and biologic stratigraphy or biostratigraphy.

The Permian through Jurassic strata of the Colorado Plateau area of southeastern Utah demonstrate the principles of stratigraphy.

Historical Development

Nicholas Steno established the theoretical basis for stratigraphy when he introduced the law of superposition, the principle of original horizontality and the principle of lateral continuity in a 1669 work on the fossilization of organic remains in layers of sediment.

The first practical large-scale application of stratigraphy was by William Smith in the 1790s and early 19th century. Smith, known as the "Father of English Geology", created the first geologic map of England and first recognized the significance of strata or rock layering and the importance of fossil markers for correlating strata. Another influential application of stratigraphy in the early 19th century was a study by Georges Cuvier and Alexandre Brongniart of the geology of the region around Paris.

Strata in Cafayate (Argentina)

Lithostratigraphy

Chalk layers in Cyprus, showing sedimentary layering

Variation in rock units, most obviously displayed as visible layering, is due to physical contrasts in rock type (lithology). This variation can occur vertically as layering (bedding), or laterally, and reflects changes in environments of deposition (known as facies change). These variations provide a lithostratigraphy or lithologic stratigraphy of the rock unit. Key concepts in stratigraphy involve understanding how certain geometric relationships between rock layers arise and what these geometries imply about their original depositional environment. The basic concept in stratigraphy, called the law of superposition, states: in an undeformed stratigraphic sequence, the oldest strata occur at the base of the sequence.

Chemostratigraphy studies the changes in the relative proportions of trace elements and isotopes within and between lithologic units. Carbon and oxygen isotope ratios vary with time, and researchers can use those to map subtle changes that occurred in the paleoenvironment. This has led to the specialized field of isotopic stratigraphy.

Cyclostratigraphy documents the often cyclic changes in the relative proportions of minerals (particularly carbonates), grain size, thickness of sediment layers (varves) and fossil diversity with time, related to seasonal or longer term changes in palaeoclimates.

Biostratigraphy

Biostratigraphy or paleontologic stratigraphy is based on fossil evidence in the rock layers. Strata from widespread locations containing the same fossil fauna and flora are said to be correlatable in time. Biologic stratigraphy was based on William Smith's principle of faunal succession, which predated, and was one of the first and most powerful lines of evidence for, biological evolution. It provides strong evidence for the formation (speciation) and extinction of species. The geologic time scale was developed during the 19th century, based on the evidence of biologic stratigraphy and faunal succession. This timescale remained a relative scale until the development of radiometric dating, which gave it and the stratigraphy it was based on an absolute time framework, leading to the development of chronostratigraphy.

One important development is the Vail curve, which attempts to define a global historical sea-level curve according to inferences from worldwide stratigraphic patterns. Stratigraphy is also commonly used to delineate the nature and extent of hydrocarbon-bearing reservoir rocks, seals, and traps of petroleum geology.

Chronostratigraphy

Chronostratigraphy is the branch of stratigraphy that places an absolute age, rather than a relative age on rock strata. The branch is concerned with deriving geochronological data for rock units, both directly and inferentially, so that a sequence of time-relative events that created the rocks formation can be derived. The ultimate aim of chronostratigraphy is to place dates on the sequence of deposition of all rocks within a geological region, and then to every region, and by extension to provide an entire geologic record of the Earth.

A gap or missing strata in the geological record of an area is called a stratigraphic hiatus. This may be the result of a halt in the deposition of sediment. Alternatively, the gap may be due to removal by erosion, in which case it may be called a stratigraphic vacuity. It is called a *hiatus* because deposition was *on hold* for a period of time. A physical gap may represent both a period of non-deposition and a period of erosion. A geologic fault may cause the appearance of a hiatus.

Magnetostratigraphy

Magnetostratigraphy is a chronostratigraphic technique used to date sedimentary and volcanic sequences. The method works by collecting oriented samples at measured intervals throughout a section. The samples are analyzed to determine their detrital remanent magnetism (DRM), that is, the polarity of Earth's magnetic field at the time a stratum was deposited. For sedimentary rocks, this is possible because, when very fine-grained magnetic minerals (< 17 μm), behave like tiny compasses as they fall through the water column and orient themselves with Earth's magnetic field. Upon burial, that orientation is preserved. For volcanic rocks, magnetic minerals, which form in the melt, orient themselves with the ambient magnetic field, and are fixed in place upon crystallization of the lava.

Oriented paleomagnetic core samples are collected in the field; mudstones, siltstones, and very fine-grained sandstones are the preferred lithologies because the magnetic grains are finer and more likely to orient with the ambient field during deposition. If the ancient magnetic field were oriented similar to today's field (North Magnetic Pole near the North Rotational Pole), the strata would retain a normal polarity. If the data indicate that the North Magnetic Pole were near the South Rotational Pole, the strata would exhibit reversed polarity.

Results of the individual samples are analyzed by removing the natural remanent magnetization (NRM) to reveal the DRM. Following statistical analysis, the results are used to generate a local magnetostratigraphic column that can then be compared against the Global Magnetic Polarity Time Scale. This technique is used to date sequences that generally lack fossils or interbedded igneous rocks. The continuous nature of the sampling means that it is also a powerful technique for the estimation of sediment-accumulation rates.

Archaeological Stratigraphy

In the field of archaeology, soil stratigraphy is used to better understand the processes that form and protect archaeological sites. Since the law of superposition holds true, it can help date finds or features from each context; these finds and features can be placed in sequence and the dates interpolated. Phases of activity can also often be seen through stratigraphy, especially when a trench or feature is viewed in section (profile). Because pits and other features can be dug down into earlier levels, not all material at the same absolute depth is necessarily of the same age; close attention has to be paid to the archeological layers. The Harris-matrix is a tool to depict complex stratigraphic relations when they are found, for example, in the context of urban archaeology.

GEOLOGICAL HISTORY OF EARTH

The **geological history of Earth** follows the major events in Earth's past based on the geologic time scale, a system of chronological measurement based on the study of the planet's rock layers (stratigraphy). Earth formed about 4.54 billion years ago by accretion from the solar nebula, a disk-shaped mass of dust and gas left over from the formation of the Sun, which also created the rest of the Solar System.

Earth was initially molten due to extreme volcanism and frequent collisions with other bodies. Eventually, the outer layer of the planet cooled to form a solid crust when water began accumulating in the atmosphere. The Moon formed soon afterwards, possibly as a result of the impact of a planetoid with the Earth. Outgassing and volcanic activity produced the primordial atmosphere. Condensing water vapor, augmented by ice delivered from comets, produced the oceans.

As the surface continually reshaped itself over hundreds of millions of years, continents formed and broke apart. They migrated across the surface, occasionally combining to form a supercontinent. Roughly 750 million years ago, the earliest-known supercontinent Rodinia, began to break apart. The continents later recombined to form Pannotia, 600 to 540 million years ago, then finally Pangaea, which broke apart 200 million years ago.

The present pattern of ice ages began about 40 million years ago, then intensified at the end of the Pliocene. The polar regions have since undergone repeated cycles of glaciation and thaw, repeating every 40,000–100,000 years. The last glacial period of the current ice age ended about 10,000 years ago.

Precambrian

The Precambrian includes approximately 90% of geologic time. It extends from 4.6 billion years ago to the beginning of the Cambrian Period (about 541 Ma). It includes three eons, the Hadean, Archean, and Proterozoic.

Hadean Eon

During Hadean time (4.6–4 Ga), the Solar System was forming, probably within a large cloud of gas and dust around the sun, called an accretion disc from which Earth formed 4,500 million years ago.

Artist's conception of a protoplanetary disc

The Hadean Eon is not formally recognized, but it essentially marks the era before we have adequate record of significant solid rocks. The oldest dated zircons date from about 4,400 million years ago.

Earth was initially molten due to extreme volcanism and frequent collisions with other bodies. Eventually, the outer layer of the planet cooled to form a solid crust when water began accumulating in the atmosphere. The Moon formed soon afterwards, possibly as a result of the impact of a large planetoid with the Earth. Some of this object's mass merged with the Earth, significantly altering its internal composition, and a portion was ejected into space. Some of the material survived to form an orbiting moon. More recent potassium isotopic studies suggest that the Moon was formed by a smaller, high-energy, high-angular-momentum giant impact cleaving off a significant portion of the Earth. Outgassing and volcanic activity produced the primordial atmosphere. Condensing water vapor, augmented by ice delivered from comets, produced the oceans.

During the Hadean the Late Heavy Bombardment occurred (approximately 4,100 to 3,800 million years ago) during which a large number of impact craters are believed to have formed on the Moon, and by inference on Earth, Mercury, Venus and Mars as well.

Archean Eon

The Earth of the early Archean (4,000 to 2,500 million years ago) may have had a different tectonic style. During this time, the Earth's crust cooled enough that rocks and continental plates began to form. Some scientists think because the Earth was hotter, that plate tectonic activity was more vigorous than it is today, resulting in a much greater rate of recycling of crustal material. This may have prevented cratonisation and continent formation until the mantle cooled and convection slowed down. Others argue that the subcontinental lithospheric mantle is too buoyant to subduct and that the lack of Archean rocks is a function of erosion and subsequent tectonic events. In contrast to the Proterozoic, Archean rocks are often heavily metamorphized deep-water sediments, such as graywackes, mudstones, volcanic sediments and banded iron formations. Greenstone belts are typical Archean formations, consisting of alternating high- and low-grade metamorphic rocks. The high-grade rocks were derived from volcanic island arcs, while the low-grade metamorphic rocks represent deep-sea sediments eroded from the neighboring island frogs and deposited in a forearc basin. In short, greenstone belts represent sutured protocontinents.

The Earth's magnetic field was established 3.5 billion years ago. The solar wind flux was about 100 times the value of the modern Sun, so the presence of the magnetic field helped prevent the planet's atmosphere from being stripped away, which is what likely happened to the atmosphere of Mars. However, the field strength was lower than at present and the magnetosphere was about half the modern radius.

Proterozoic Eon

The geologic record of the Proterozoic (2,500 to 541 million years ago) is more complete than that for the preceding Archean. In contrast to the deep-water deposits of the Archean, the Proterozoic features many strata that were laid down in extensive shallow epicontinental seas; furthermore, many of these rocks are less metamorphosed than Archean-age ones, and plenty are unaltered. Study of these rocks show that the eon featured massive, rapid continental accretion (unique to the Proterozoic), supercontinent cycles, and wholly modern orogenic activity. Roughly 750 million years ago, the earliest-known supercontinent Rodinia, began to break apart. The continents later recombined to form Pannotia, 600–540 Ma.

The first-known glaciations occurred during the Proterozoic, one began shortly after the beginning of the eon, while there were at least four during the Neoproterozoic, climaxing with the Snowball Earth of the Varangian glaciation.

Phanerozoic Eon

The **Phanerozoic** Eon is the current eon in the geologic timescale. It covers roughly 541 million years. During this period continents drifted about, eventually collected into a single landmass known as Pangea and then split up into the current continental landmasses.

The Phanerozoic is divided into three eras — the Paleozoic, the Mesozoic and the Cenozoic.

Paleozoic Era

The **Paleozoic** spanned from roughly 541 to 252 million years ago (Ma) and is subdivided into six geologic periods; from oldest to youngest they are the Cambrian, Ordovician, Silurian, Devonian, Carboniferous and Permian. Geologically, the Paleozoic starts shortly after the breakup of a supercontinent called Pannotia and at the end of a global ice age. Throughout the early Paleozoic, the Earth's landmass was broken up into a substantial number of relatively small continents. Toward the end of the era the continents gathered together into a supercontinent called Pangaea, which included most of the Earth's land area.

Cambrian Period

The **Cambrian** is a major division of the geologic timescale that begins about 541.0 ± 1.0 Ma. Cambrian continents are thought to have resulted from the breakup of a Neoproterozoic supercontinent called Pannotia. The waters of the Cambrian period appear to have been widespread and shallow. Continental drift rates may have been anomalously high. Laurentia, Baltica and Siberia remained independent continents following the break-up of the supercontinent of Pannotia. Gondwana started to drift toward the South Pole. Panthalassa covered most of the southern hemisphere, and minor oceans included the Proto-Tethys Ocean, Iapetus Ocean and Khanty Ocean.

Ordovician Period

The **Ordovician** Period started at a major extinction event called the Cambrian-Ordovician extinction events some time about 485.4 ± 1.9 Ma. During the Ordovician the southern continents were collected into a single continent called Gondwana. Gondwana started the period in the equatorial latitudes and, as the period progressed, drifted toward the South Pole. Early in the Ordovician the continents Laurentia, Siberia and Baltica were still independent continents (since the break-up of the supercontinent Pannotia earlier), but Baltica began to move toward Laurentia later in the period, causing the Iapetus Ocean to shrink between them. Also, Avalonia broke free from Gondwana and began to head north toward Laurentia. The Rheic Ocean was formed as a result of this. By the end of the period, Gondwana had neared or approached the pole and was largely glaciated.

The Ordovician came to a close in a series of extinction events that, taken together, comprise the second-largest of the five major extinction events in Earth's history in terms of percentage of genera that became extinct. The only larger one was the Permian-Triassic extinction event. The extinctions occurred approximately 447 to 444 million years ago and mark the boundary between the Ordovician and the following Silurian Period.

The most-commonly accepted theory is that these events were triggered by the onset of an ice age, in the Hirnantian faunal stage that ended the long, stable greenhouse conditions typical of the Ordovician. The ice age was probably not as long-lasting as once thought; study of oxygen isotopes in fossil brachiopods shows that it was probably no longer than 0.5 to 1.5 million years. The event was preceded by a fall in atmospheric carbon dioxide (from 7000ppm to 4400ppm) which selectively affected the shallow seas where most organisms lived. As the southern supercontinent Gondwana drifted over the South Pole, ice caps formed on it. Evidence of these ice caps have been detected in Upper Ordovician rock strata of North Africa and then-adjacent northeastern South America, which were south-polar locations at the time.

Silurian Period

The **Silurian** is a major division of the geologic timescale that started about 443.8 ± 1.5 Ma. During the Silurian, Gondwana continued a slow southward drift to high southern latitudes, but there is evidence that the Silurian ice caps were less extensive than those of the late Ordovician glaciation. The melting of ice caps and glaciers contributed to a rise in sea levels, recognizable from the fact that Silurian sediments overlie eroded Ordovician sediments, forming an unconformity. Other cratons and continent fragments drifted together near the equator, starting the formation of a second supercontinent known as Euramerica. The vast ocean of Panthalassa covered most of the northern hemisphere. Other minor oceans include Proto-Tethys, Paleo-Tethys,

Rheic Ocean, a seaway of Iapetus Ocean (now in between Avalonia and Laurentia), and newly formed Ural Ocean.

Devonian Period

The **Devonian** spanned roughly from 419 to 359 Ma. The period was a time of great tectonic activity, as Laurasia and Gondwana drew closer together. The continent Euramerica (or Laurussia) was created in the early Devonian by the collision of Laurentia and Baltica, which rotated into the natural dry zone along the Tropic of Capricorn. In these near-deserts, the Old Red Sandstone sedimentary beds formed, made red by the oxidized iron (hematite) characteristic of drought conditions. Near the equator Pangaea began to consolidate from the plates containing North America and Europe, further raising the northern Appalachian Mountains and forming the Caledonian Mountains in Great Britain and Scandinavia. The southern continents remained tied together in the supercontinent of Gondwana. The remainder of modern Eurasia lay in the Northern Hemisphere. Sea levels were high worldwide, and much of the land lay submerged under shallow seas. The deep, enormous Panthalassa (the universal ocean) covered the rest of the planet. Other minor oceans were Paleo-Tethys, Proto-Tethys, Rheic Ocean and Ural Ocean (which was closed during the collision with Siberia and Baltica).

Carboniferous Period

The **Carboniferous** extends from about 358.9 ± 0.4 to about 298.9 ± 0.15 Ma.

A global drop in sea level at the end of the Devonian reversed early in the Carboniferous; this created the widespread epicontinental seas and carbonate deposition of the Mississippian. There was also a drop in south polar temperatures; southern Gondwana was glaciated throughout the period, though it is uncertain if the ice sheets were a holdover from the Devonian or not. These conditions apparently had little effect in the deep tropics, where lush coal swamps flourished within 30 degrees of the northernmost glaciers. A mid-Carboniferous drop in sea-level precipitated a major marine extinction, one that hit crinoids and ammonites especially hard. This sea-level drop and the associated unconformity in North America separate the Mississippian Period from the Pennsylvanian period.

The Carboniferous was a time of active mountain building, as the supercontinent Pangea came together. The southern continents remained tied together in the supercontinent Gondwana, which collided with North America-Europe (Laurussia) along the present line of eastern North America. This continental collision resulted in the Hercynian orogeny in Europe, and the Alleghenian orogeny in North America; it also extended the newly uplifted Appalachians southwestward as the Ouachita Mountains. In the same time frame, much of present eastern Eurasian plate welded itself to Europe along the line of the Ural mountains. There were two major oceans in the Carboniferous the Panthalassa and Paleo-Tethys. Other minor oceans were shrinking and eventually closed the Rheic Ocean (closed by the assembly of South and North America), the small, shallow Ural Ocean (which was closed by the collision of Baltica, and Siberia continents, creating the Ural Mountains) and Proto-Tethys Ocean.

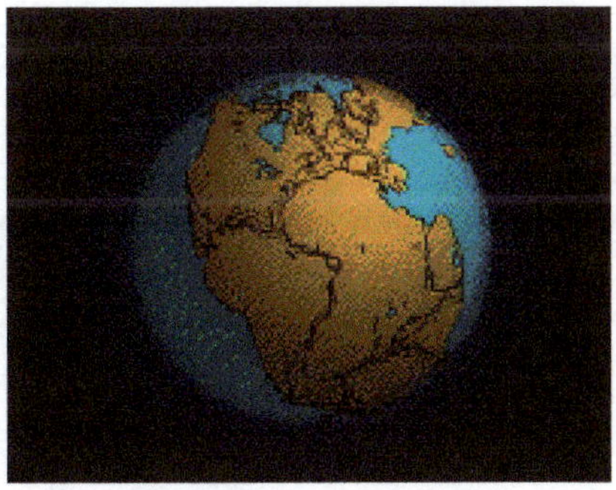

Pangaea separation animation

Permian Period

The **Permian** extends from about 298.9 ± 0.15 to 252.17 ± 0.06 Ma.

During the Permian all the Earth's major land masses, except portions of East Asia, were collected into a single supercontinent known as Pangaea. Pangaea straddled the equator and extended toward the poles, with a corresponding effect on ocean currents in the single great ocean (*Panthalassa*, the *universal sea*), and the Paleo-Tethys Ocean, a large ocean that was between Asia and Gondwana. The Cimmeria continent rifted away from Gondwana and drifted north to Laurasia, causing the Paleo-Tethys to shrink. A new ocean was growing on its southern end, the Tethys Ocean, an ocean that would dominate much of the Mesozoic Era. Large continental landmasses create climates with extreme variations of heat and cold ("continental climate") and monsoon conditions with highly seasonal rainfall patterns. Deserts seem to have been widespread on Pangaea.

Mesozoic Era

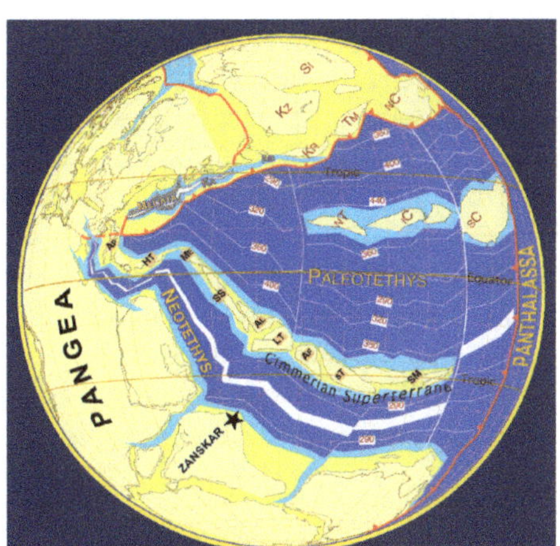

Plate tectonics- 249 million years ago

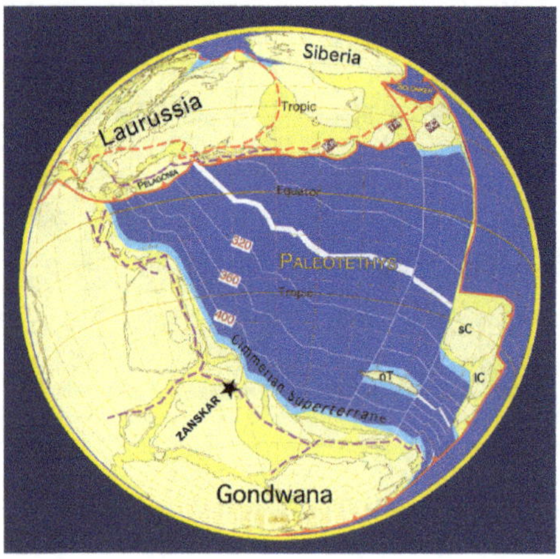

Plate tectonics- 290 million years ago

The **Mesozoic** extended roughly from 252 to 66 million years ago.

After the vigorous convergent plate mountain-building of the late Paleozoic, Mesozoic tectonic deformation was comparatively mild. Nevertheless, the era featured the dramatic rifting of the supercontinent Pangaea. Pangaea gradually split into a northern continent, Laurasia, and a southern continent, Gondwana. This created the passive continental margin that characterizes most of the Atlantic coastline (such as along the U.S. East Coast) today.

Triassic Period

The **Triassic** Period extends from about 252.17 ± 0.06 to 201.3 ± 0.2 Ma. During the Triassic, almost all the Earth's land mass was concentrated into a single supercontinent centered more or less on the equator, called Pangaea ("all the land"). This took the form of a giant "Pac-Man" with an east-facing "mouth" constituting the Tethys sea, a vast gulf that opened farther westward in the mid-Triassic, at the expense of the shrinking Paleo-Tethys Ocean, an ocean that existed during the Paleozoic.

The remainder was the world-ocean known as Panthalassa (all the sea). All the deep-ocean sediments laid down during the Triassic have disappeared through subduction of oceanic plates; thus, very little is known of the Triassic open ocean. The supercontinent Pangaea was rifting during the Triassic—especially late in the period—but had not yet separated. The first nonmarine sediments in the rift that marks the initial break-up of Pangea—which separated New Jersey from Morocco—are of Late Triassic age; in the U.S., these thick sediments comprise the Newark Supergroup. Because of the limited shoreline of one super-continental mass, Triassic marine deposits are globally relatively rare; despite their prominence in Western Europe, where the Triassic was first studied. In North America, for example, marine deposits are limited to a few exposures in the west. Thus Triassic stratigraphy is mostly based on organisms living in lagoons and hypersaline environments, such as *Estheria* crustaceans and terrestrial vertebrates.

Jurassic Period

The **Jurassic** Period extends from about 201.3 ± 0.2 to 145.0 Ma. During the early Jurassic, the supercontinent Pangaea broke up into the northern supercontinent Laurasia and the southern supercontinent Gondwana; the Gulf of Mexico opened in the new rift between North America and what is now Mexico's Yucatan Peninsula. The Jurassic North Atlantic Ocean was relatively narrow, while the South Atlantic did not open until the following Cretaceous Period, when Gondwana itself rifted apart. The Tethys Sea closed, and the Neotethys basin appeared. Climates were warm, with no evidence of glaciation. As in the Triassic, there was apparently no land near either pole, and no extensive ice caps existed. The Jurassic geological record is good in Western Europe, where extensive marine sequences indicate a time when much of the continent was submerged under shallow tropical seas; famous locales include the Jurassic Coast World Heritage Site and the renowned late Jurassic *lagerstätten* of Holzmaden and Solnhofen. In contrast, the North American Jurassic record is the poorest of the Mesozoic, with few outcrops at the surface. Though the epicontinental Sundance Sea left marine deposits in parts of the northern plains of the United States and Canada during the late Jurassic, most exposed sediments from this period are continental, such as the alluvial deposits of the Morrison Formation. The first of several massive batholiths were emplaced in the northern Cordillera beginning in the mid-Jurassic, marking the Nevadan orogeny. Important Jurassic exposures are also found in Russia, India, South America, Japan, Australasia and the United Kingdom.

Cretaceous Period

Plate tectonics- 100 Ma,Cretaceous period

The **Cretaceous** Period extends from circa 145 million years ago to 66 million years ago.

During the Cretaceous, the late Paleozoic-early Mesozoic supercontinent of Pangaea completed its breakup into present day continents, although their positions were substantially different at the time. As the Atlantic Ocean widened, the convergent-margin orogenies that had begun during the Jurassic continued in the North American Cordillera, as the Nevadan orogeny was followed by the Sevier and Laramide orogenies. Though Gondwana was still intact in the beginning of the Cretaceous, Gondwana itself broke up as South America, Antarctica and Australia rifted away from Africa (though India and Madagascar remained attached to each other); thus, the South Atlantic and Indian Oceans were newly formed. Such active rifting lifted great undersea mountain chains along the welts, raising eustatic sea levels worldwide.

To the north of Africa the Tethys Sea continued to narrow. Broad shallow seas advanced across central North America (the Western Interior Seaway) and Europe, then receded late in the period, leaving thick marine deposits sandwiched between coal beds. At the peak of the Cretaceous transgression, one-third of Earth's present land area was submerged. The Cretaceous is justly famous for its chalk; indeed, more chalk formed in the Cretaceous than in any other period in the Phanerozoic. Mid-ocean ridge activity—or rather, the circulation of seawater through the enlarged ridges—enriched the oceans in calcium; this made the oceans more saturated, as well as increased the bioavailability of the element for calcareous nanoplankton. These widespread carbonates and other sedimentary deposits make the Cretaceous rock record especially fine. Famous formations from North America include the rich marine fossils of Kansas's Smoky Hill Chalk Member and the terrestrial fauna of the late Cretaceous Hell Creek Formation. Other important Cretaceous exposures occur in Europe and China. In the area that is now India, massive lava beds called the Deccan Traps were laid down in the very late Cretaceous and early Paleocene.

Cenozoic Era

The **Cenozoic** Era covers the 66 million years since the Cretaceous–Paleogene extinction event up to and including the present day. By the end of the Mesozoic Era, the continents had rifted into nearly their present form. Laurasia became North America and Eurasia, while Gondwana split into South America, Africa, Australia, Antarctica and the Indian subcontinent, which collided with the Asian plate. This impact gave rise to the Himalayas. The Tethys Sea, which had separated the northern continents from Africa and India, began to close up, forming the Mediterranean sea.

Paleogene Period

The **Paleogene** (alternatively **Palaeogene**) Period is a unit of geologic time that began 66 and ended 23.03 Ma and comprises the first part of the Cenozoic Era. This period consists of the Paleocene, Eocene and Oligocene Epochs.

Paleocene Epoch

The **Paleocene**, lasted from 66 million years ago to 56 million years ago.

In many ways, the Paleocene continued processes that had begun during the late Cretaceous Period. During the Paleocene, the continents continued to drift toward their present positions. Supercontinent Laurasia had not yet separated into three continents. Europe and Greenland were still connected. North America and Asia were still intermittently joined by a land bridge, while Greenland and North America were beginning to separate. The Laramide orogeny of the late Cretaceous continued to uplift the Rocky Mountains in the American west, which ended in the succeeding epoch. South and North America remained separated by equatorial seas (they joined during the Neogene); the components of the former southern supercontinent Gondwana continued to split apart, with Africa, South America, Antarctica and Australia pulling away from each other. Africa was heading north toward Europe, slowly closing the Tethys Ocean, and India began its migration to Asia that would lead to a tectonic collision and the formation of the Himalayas.

Eocene Epoch

During the **Eocene** (56 million years ago - 33.9 million years ago), the continents continued to drift toward their present positions. At the beginning of the period, Australia and Antarctica remained connected, and warm equatorial currents mixed with colder Antarctic waters, distributing the heat around the world and keeping global temperatures high. But when Australia split from the southern continent around 45 Ma, the warm equatorial currents were deflected away from Antarctica, and an isolated cold water channel developed between the two continents. The Antarctic region cooled down, and the ocean surrounding Antarctica began to freeze, sending cold water and ice floes north, reinforcing the cooling. The present pattern of ice ages began about 40 million years ago.

The northern supercontinent of Laurasia began to break up, as Europe, Greenland and North America drifted apart. In western North America, mountain building started in the Eocene, and huge lakes formed in the high flat basins among uplifts. In Europe, the Tethys Sea finally vanished, while the uplift of the Alps isolated its final remnant, the Mediterranean, and created another shallow sea with island archipelagos to the north. Though the North Atlantic was opening, a land connection appears to have remained between North America and Europe since the faunas of the two regions are very similar. India continued its journey away from Africa and began its collision with Asia, creating the Himalayan orogeny.

Oligocene Epoch

The **Oligocene** Epoch extends from about 34 million years ago to 23 million years ago. During the Oligocene the continents continued to drift toward their present positions.

Antarctica continued to become more isolated and finally developed a permanent ice cap. Mountain building in western North America continued, and the Alps started to rise in Europe as the African plate continued to push north into the Eurasian plate, isolating the remnants of Tethys Sea. A brief marine incursion marks the early Oligocene in Europe. There appears to have been a land bridge in the early Oligocene between North America and Europe since the faunas of the two regions are very similar. During the Oligocene, South America was finally detached from Antarctica and drifted north toward North America. It also allowed the Antarctic Circumpolar Current to flow, rapidly cooling the continent.

Neogene Period

The **Neogene** Period is a unit of geologic time starting 23.03 Ma. and ends at 2.588 Mya. The Neogene Period follows the Paleogene Period. The Neogene consists of the Miocene and Pliocene and is followed by the Quaternary Period.

Miocene Epoch

The **Miocene** extends from about 23.03 to 5.333 Ma.

During the Miocene continents continued to drift toward their present positions. Of the modern geologic features, only the land bridge between South America and North America was absent, the subduction zone along the Pacific Ocean margin of South America caused the rise of the Andes and the southward extension of the Meso-American peninsula. India continued to collide with Asia. The Tethys Seaway continued to shrink and then disappeared as Africa collided with Eurasia in the Turkish-Arabian region between 19 and 12 Ma (ICS 2004). Subsequent uplift of mountains in the western Mediterranean region and a global fall in sea levels combined to cause a temporary drying up of the Mediterranean Sea resulting in the Messinian salinity crisis near the end of the Miocene.

Pliocene Epoch

The **Pliocene** extends from 5.333 million years ago to 2.588 million years ago. During the Pliocene continents continued to drift toward their present positions, moving from positions possibly as far as 250 kilometres (155 mi) from their present locations to positions only 70 km from their current locations.

South America became linked to North America through the Isthmus of Panama during the Pliocene, bringing a nearly complete end to South America's distinctive marsupial faunas. The formation of the Isthmus had major consequences on global temperatures, since warm equatorial ocean currents were cut off and an Atlantic cooling cycle began, with cold Arctic and Antarctic waters dropping temperatures in the now-isolated Atlantic Ocean. Africa's collision with Europe formed the Mediterranean Sea, cutting off the remnants of the Tethys Ocean. Sea level changes exposed the land-bridge between Alaska and Asia. Near the end of the Pliocene, about 2.58 million years ago (the start of the Quaternary Period), the current ice age began. The polar regions have since undergone repeated cycles of glaciation and thaw, repeating every 40,000–100,000 years.

Quaternary Period

Pleistocene Epoch

The **Pleistocene** extends from 2.588 million years ago to 11,700 years before present. The modern continents were essentially at their present positions during the Pleistocene, the plates upon which they sit probably having moved no more than 100 kilometres (62 mi) relative to each other since the beginning of the period.

Holocene Epoch

The **Holocene** Epoch began approximately 11,700 calendar years before present and continues to the present. During the Holocene, continental motions have been less than a kilometer.

The last glacial period of the current ice age ended about 10,000 years ago. Ice melt caused world sea levels to rise about 35 metres (115 ft) in the early part of the Holocene. In addition, many areas above about 40 degrees north latitude had been depressed by the weight of the Pleistocene glaciers and rose as much as 180 metres (591 ft) over the late Pleistocene and Holocene, and are still rising today. The sea level rise and temporary land depression allowed temporary marine incursions into areas that are now far from the sea. Holocene marine fossils are known from Vermont, Quebec, Ontario and Michigan. Other than higher latitude temporary marine incursions associated with glacial depression, Holocene fossils are found primarily in lakebed, floodplain and cave deposits. Holocene marine deposits along low-latitude coastlines are rare because the rise in sea levels during the period exceeds any likely upthrusting of non-glacial origin. Post-glacial rebound in Scandinavia resulted in the emergence of coastal areas around the Baltic Sea, including much of Finland. The region continues to rise, still causing weak earthquakes across Northern Europe. The equivalent event in North America was the rebound of Hudson Bay, as it shrank from its larger, immediate post-glacial Tyrrell Sea phase, to near its present boundaries.

TIMELINE OF GEOLOGY

Early works

- c. 1025 – Abū Rayhān al-Bīrūnī publishes the *Kitab fi Tahqiq ma li'l-Hind* (*Researches on India*), in which he discusses the geology of India and hypothesizes that it was once a sea.
- 1027 – Avicenna publishes *The Book of Healing*, in which he hypothesizes on two causes of mountains, and proposes the law of superposition and the concept of uniformitarianism.

16th and 17th centuries

Portuguese and Spanish explorers systematically measure magnetic declination to estimate the geographical longitude

- 1556 – Agricola publishes *De re metallica*. This book acts as the standard mining and assaying text for the next 250 years.
- 1596 - Abraham Ortelius, Flemish-Spanish cartographer, first envisages the continental drift theory.
- 1603 - Ulisse Aldrovandi coins the term *Geology*.
- 1669 – Nicolas Steno puts forward his theory that sedimentary strata had been deposited in former seas, and that fossils were organic in origin

18th Century

- 1701 – Edmund Halley suggests using the salinity and evaporation of the Mediterranean to determine the age of the Earth
- 1743 – Dr Christopher Packe produces a geological map of south-east England
- 1746 – Jean-Étienne Guettard presents the first mineralogical map of France to the French Academy of Sciences.
- 1760 – John Michell suggests earthquakes are caused by one layer of rocks rubbing against another
- 1776 – James Keir suggests that some rocks, such as those at the Giant's Causeway, might have been formed by the crystallisation of molten lava

- 1779 – Comte de Buffon speculates that the Earth is older than the 6,000 years suggested by the Bible
- 1785 – James Hutton presents paper entitled Theory of the Earth – earth must be old
- 1799 – William Smith produces the first large scale geological map, of the area around Bath

19th Century

- 1809 – William Maclure conducts the first geological survey of the eastern United States
- 1830 – Sir Charles Lyell publishes book, Principles of Geology, which describes the world as being several hundred million years old
- 1837 – Louis Agassiz begins his glaciation studies which eventually demonstrate that the Earth has had at least one ice age
- 1841 – August Breithaupt, *Vollstandiges Handbuch der Mineralogie*
- 1848 – James Dwight Dana, *Manual of Mineralogy*
- 1862 – Lord Kelvin attempts to find the age of the Earth by examining its cooling time and estimates that the Earth is between 20—400 million years old
- 1884 – Marcel Alexandre Bertrand, Nappe and Thrust fault theory

20th Century

- 1903 – George Darwin and John Joly claim that radioactivity is partially responsible for the Earth's heat
- 1907 – Bertram Boltwood proposes that the amount of lead in uranium and thorium ores might be used to determine the Earth's age and crudely dates some rocks to have ages between 410—2200 million years
- 1911 – Arthur Holmes uses radioactivity to date rocks, the oldest being 1.6 billion years old
- 1912 – Alfred Wegener proposes that all the continents once formed a single landmass called Pangaea that broke apart via continental drift
- 1912 - George Barrow maps zones of metamorphism (the Barrovian sequence) in southern Scotland
- 1913 – Albert A. Michelson measures tides in the solid body of the Earth
- 1915 - Pentti Eskola develops the concept of metamorphic facies
- 1928 – N. L. Bowen publishes *The Evolution of the Igneous Rocks,* revolutionizing experimental igneous petrology
- 1935 – Charles Richter invents a logarithmic scale to measure the intensity of earthquakes ({\displaystyle M_{L}})
- 1941 – Nickel-Strunz classification, Karl H. Strunz, *Mineralogische Tabellen*
- (1948 - 1959) – Felix Andries Vening Meinesz investigations show gravity anomalies, implying that the crust is moving (together with J.H.F. Umbgrove, B.G. Escher and Ph.H. Kuenen)
- 1951 – Alfred Rittmann links subduction, volcanism and the Wadati-Benioff zone

- 1953 – Maurice Ewing and Bruce Heezen discover the Great Global Rift running along the Mid-Atlantic Ridge
- 1960 – Harry Hess proposes that new sea floor might be created at mid-ocean rifts and destroyed at deep sea trenches
- 1963 – Frederick Vine and Drummond Matthews explain the stripes of magnetized rocks with alternating magnetic polarities running parallel to mid- ocean ridges as due to sea floor spreading and the periodic geomagnetic field reversals (Vine–Matthews–Morley hypothesis)
- 1966 – Keiiti Aki discovers the seismic moment ({\displaystyle M_{0}})
- 1979 – Thomas C. Hanks and Hiroo Kanamori, Moment magnitude scale ({\displaystyle M_{W}}), it succeeds the Richter magnitude scale
- 1980 – Physicist Luis Alvarez, his son, geologist Walter Alvarez, and others propose that the impact of a large extraterrestrial object caused the extinction of the dinosaurs at the end of the Cretaceous Period, about 66 million years ago.

21st century

- 2001 – Nickel-Strunz classification, Karl H. Strunz and Ernest H. Nickel, *Strunz Mineralogical Tables* 9 ed.

REFERENCES

1. Gohau 1990, p. 7
2. Moore, Ruth. The Earth We Live On. New York: Alfred A. Knopf, 1956. p. 13
3. Aristotle. Meteorology. Book 1, Part 14
4. Asimov, M. S.; Bosworth, Clifford Edmund (eds.). The Age of Achievement: A.D. 750 to the End of the Fifteenth Century : The Achievements. History of civilizations of Central Asia. pp. 211–214. ISBN 978-92-3-102719-2.
5. Frank 1938, p. 96
6. Gohau 1990, p. 118
7. Gohau 1990, p. 114
8. Kooijmans 2007
9. Wyse Jackson 2007
10. Woods 2005, pp. 4 & 96
11. Jardine, Secord & Spary 1996, pp. 212–214
12. Gohau 1990, p. 88
13. Gohau 1990, p. 92
14. Jardine, Secord & Spary 1996, p. 232
15. Gohau 1990, p. 8
16. Gohau 1990, p. 219
17. Eddy, Matthew Daniel (2008). The Language of Mineralogy: John Walker, Chemistry and the Edinburgh Medical School. Ashgate.

18. Frank, Adams Dawson. The Birth and Development of the Geological Sciences. Baltimore: The Williams & Wilkins Company, 1938. p. 209

19. Albritton, Claude C. The Abyss of Time. San Francisco: Freeman, Cooper & Company, 1980. p. 95-96

20. Frank 1938, p. 239

21. Albritton, Claude C. The Abyss of Time. San Francisco: Freeman, Cooper & Company, 1980. p. 104-107

22. Bowler 1992, p. 216

23. Gohau 1990, p. 144

24. Second J A (1986) Controversy in Victorian Geology: The Cambrian-Silurian Dispute Princeton University Press, 301pp, ISBN 0-691-02441-3

25. Herbert, Sandra. Charles Darwin as a prospective geological author, British Journal for the History of Science 24. 1991. p. 159–192

26. Gohau 1990, p. 145

27. Frank 1938, p. 226

28. Keynes, Richard ed.. Charles Darwin's zoology notes & specimen lists from H.M.S. Beagle, Cambridge University Press, 2000. p. ix

29. Dalrymple, G. Brent (1994). The Age of the Earth. Stanford University Press. p. 74. ISBN 0-8047-2331-1.

30. England, P; Molnar, P (2007). John Perry's neglected critique of Kelvin's age for the Earth: A missed opportunity in geodynamics. GSA Today. 17 (1).

31. Dalrymple, G. Brent (1994). The Age Of The Earth. Stanford University Press. pp. 14–17, 38. ISBN 0-8047-2331-1.

32. Holmes, Arthur (1913). The Age of the Earth. London: Harper. p. 18.

33. Dalrymple, G.Brent (2004). Ancient Earth, Ancient Skies: The Age of Earth and Its Cosmic Surroundings. Stanford University Press. p. 52. ISBN 0804749337.

34. Dalrymple 2004, p. 156

35. Wegener, Alfred (1912). Die Herausbildung der Grossformen der Erdrinde (Kontinente und Ozeane), auf geophysikalischer Grundlage (PDF). Petermanns Geographische Mitteilungen,. 63: 185–195, 253–256, 305–309.

36. Holmes, Arthur (1931). Radioactivity and Earth Movements (PDF). Transactions of the Geological Society of Glasgow. Geological Society of Glasgow: 559–606.

37. Hess, H. H. (November 1, 1962). History of Ocean Basins (PDF). In A. E. J. Engel; Harold L. James; B. F. Leonard. Petrologic studies: a volume in honor of A. F. Buddington. Boulder, CO: Geological Society of America. pp. 599–620.

38. Wilson, J.T> (1963). Hypothesis on the Earth's behaviour. Nature. 198 *(4884): 849–865.*

39. Wilson, J. Tuzo (1965). A new class of faults and their bearing on continental drift. Nature. 207 *(4995): 343–347.*

40. Blacket, P.M.S.; Bullard, E.; Runcorn, S.K., eds. (1965). A Symposium on Continental Drift, held on 28 October 1965. Philosophical Transactions of the Royal Society A. 258 *(1088). Royal Society of London.*

41. Studying Earth Sciences. British Geological Survey. 2006. Natural Environment Research Council. http://www.bgs.ac.uk/vacancies/studying.htm , accessed 29 November 2006

42. Rocchio, Laura. The Landsat Program. National Aeronautics and Space Administration. http://landsat.gsfc.nasa.gov , accessed 4 December 2006

CHAPTER 3

Principles of Geology

Chapter Contents

- **Cross-Cutting Relationships**
- **Types**
- **Scale**
- **Other Use**
- **Law of Included Fragments**
- **Uniformitarianism**
- **History**
- **Principle of Original Horizontality**
- **Law of Superposition**
- **Principle of Faunal Succession**

Principles of Geology: being an attempt to explain the former changes of the Earth's surface, by reference to causes now in operation is a book by the Scottish geologist Charles Lyell, first published in 3 volumes in 1830–1833.Lyell used geology throughout as a basis to strengthen his argument for Uniformitarianism. He used geological proof to determine that the Earth was older than 6,000 years, as had been previously contested. The book shows that the processes that are occurring in the present are the same processes that occurred in the past.The book was influential, not least on the young graduate Charles Darwin. The central argument in Principles was that "the present is the key to the past": that geological remains from the distant past could, and should, be explained by reference to geological processes now in operation and thus directly observable.The book is notable for being one of the first to use the term "evolution" in the context of biological speciation.

CROSS-CUTTING RELATIONSHIPS

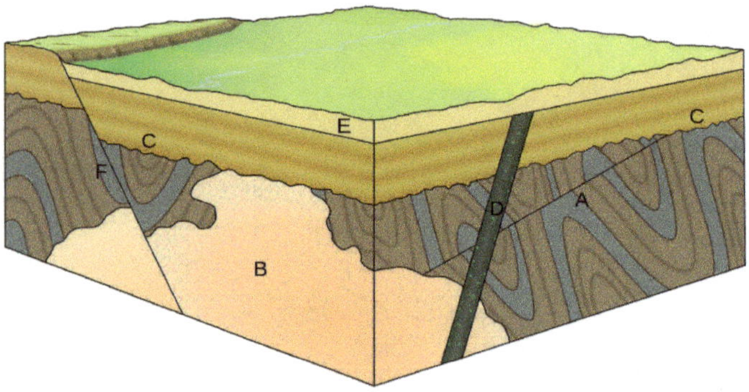

Cross-cutting relations can be used to determine the relative ages of rock strata and other geological structures. Explanations: A - folded rock strata cut by a thrust fault; B - large intrusion (cutting through A); C - erosional angular unconformity (cutting off A & B) on which rock strata were deposited; D - volcanic dike (cutting through A, B & C); E - even younger rock strata (overlying C & D); F - normal fault (cutting through A, B, C & possibly E).

Cross-cutting relationships is a principle of geology that states that the geologic feature which cuts another is the younger of the two features. It is a relative dating technique in geology. It was first developed by Danish geological pioneer Nicholas Steno in *Dissertationis prodromus* (1669) and later formulated by James Hutton in *Theory of the Earth* (1795) and embellished upon by Charles Lyell in *Principles of Geology* (1830).

TYPES

There are several basic types of cross cutting relationships:

- Structural relationships may be faults or fractures cutting through an older rock.
- Intrusional relationships occur when an igneous pluton or dike is intruded into pre-existing rocks.
- Stratigraphic relationships may be an erosional surface (or unconformity) cuts across older rock layers, geological structures, or other geological features.
- Sedimentological relationships occur where currents have eroded or scoured older sediment in a local area to produce, for example, a channel filled with sand.
- Paleontological relationships occur where animal activity or plant growth produces truncation. This happens, for example, where animal burrows penetrate into pre-existing sedimentary deposits.
- Geomorphological relationships may occur where a surficial feature, such as a river, flows through a gap in a ridge of rock. In a similar example, an impact crater excavates into a subsurface layer of rock.

Cross-cutting relationships may be compound in nature. For example, if a fault were truncated by an unconformity, and that unconformity cut by a dike. Based upon such compound cross-cutting relationships it can be seen that the fault is older than the unconformity which in turn is older than the dike. Using such rationale, the sequence of geological events can be better understood.

SCALE

Cross-cutting relationships involving an andesitic dike in Peru that cuts across the lower sedimentary strata. Both the dike and the lower strata are cut by an unconformity

A light-gray igneous intrusion in Sweden cut by a younger white pegmatite dike, which in turn is cut by an even younger black diabase dike

Cross-cutting relationships may be seen cartographically, megascopically, and microscopically. In other words, these relationships have various scales. A cartographic crosscutting relationship might look like, for example, a large fault dissecting the landscape on a large map. Megascopic cross-cutting relationships are features like igneous dikes, as mentioned above, which would be seen on an outcrop or in a limited geographic area. Microscopic cross-cutting relationships are those that require study by magnification or other close scrutiny. For example, penetration of a fossil shell by the drilling action of a boring organism is an example of such a relationship.

OTHER USE

Cross-cutting relationships can also be used in conjunction with radiometric age dating to effect an age bracket for geological materials that cannot be directly dated by radiometric techniques. For example, if a layer of sediment containing a fossil of interest is bounded on the top and bottom by unconformities, where the lower unconformity truncates dike A and the upper unconformity truncates dike B (which penetrates the layer in question), this method can be used. A radiometric age date from crystals in dike A will give the maximum age date for the layer in question and likewise, crystals from dike B will give us the minimum age date. This provides an age bracket, or range of possible ages, for the layer in question.

LAW OF INCLUDED FRAGMENTS

The **law of included fragments** is a method of relative dating in geology. Essentially, this law states that clasts in a rock are older than the rock itself. One example of this is a xenolith, which is a fragment of country rock that fell into passing magma as a result of stoping. Another example is a derived fossil, which is a fossil that has been eroded from an older bed and redeposited into a younger one. This is a restatement of Charles Lyell's original **principle of inclusions and components** from his 1830 to 1833 multi-volume *Principles of Geology*, which states that, with sedimentary rocks, if inclusions (or clasts) are found in a formation, then the inclusions must be older than the formation that contains them. For example, in sedimentary rocks, it is common for gravel from an older formation to be ripped up and included in a newer layer. A similar situation with igneous rocks occurs when xenoliths are found. These foreign bodies are picked up as magma or lava flows, and are incorporated, later to cool in the matrix. As a result, xenoliths are older than the rock which contains them.

UNIFORMITARIANISM

Uniformitarianism is the assumption that the same natural laws and processes that operate in the universe now have always operated in the universe in the past and apply everywhere in the universe. It refers to invariance in the metaphysical principles underpinning science, such as the constancy of causal structure throughout space-time, but has also been used to describe spatiotemporal invariance of physical laws. Though an unprovable postulate that cannot be verified using the scientific method, uniformitarianism has been a key first principle of virtually all fields of science.

Hutton's Unconformity at Jedburgh. A photograph shows the current scene (2003), below John Clerk of Eldin's illustration of 1787.

In geology, uniformitarianism has included the gradualistic concept that "the present is the key to the past" and is functioning at the same rates, though many modern geologists no longer hold to a strict gradualism. Coined by William Whewell, it was originally proposed in contrast to catastrophism by British naturalists in the late 18th century, starting with the work of the Scottish geologist James Hutton. Hutton's work was later refined by John Playfair and popularised by Charles Lyell's *Principles of Geology* in 1830.

HISTORY

18th century

Cliff at the east of Siccar Point in Berwickshire, showing the near-horizontal red sandstone layers above vertically tilted greywacke rocks.

The earlier conceptions likely had little influence on 18th-century European geological explanations for the formation of Earth. Abraham Gottlob Werner (1749-1817) proposed Neptunism, where strata represented deposits from shrinking seas precipitated onto primordial rocks such as granite. In 1785 James Hutton proposed an opposing, self-maintaining infinite cycle based on natural history and not on the Biblical account.

The solid parts of the present land appear in general, to have been composed of the productions of the sea, and of other materials similar to those now found upon the shores. Hence we find reason to conclude:

1st, That the land on which we rest is not simple and original, but that it is a composition, and had been formed by the operation of second causes.

2nd, That before the present land was made, there had subsisted a world composed of sea and land, in which were tides and currents, with such operations at the bottom of the sea as now take place. And,

Lastly, That while the present land was forming at the bottom of the ocean, the former land maintained plants and animals; at least the sea was then inhabited by animals, in a similar manner as it is at present.

Hence we are led to conclude, that the greater part of our land, if not the whole had been produced by operations natural to this globe; but that in order to make this land a permanent body, resisting the operations of the waters, two things had been required;

1st, The consolidation of masses formed by collections of loose or incoherent materials;

2ndly, The elevation of those consolidated masses from the bottom of the sea, the place where they were collected, to the stations in which they now remain above the level of the ocean.

Hutton then sought evidence to support his idea that there must have been repeated cycles, each involving deposition on the seabed, uplift with tilting and erosion, and then moving undersea again for further layers to be deposited. At Glen Tilt in the Cairngorm mountains he found granite penetrating metamorphic schists, in a way which indicated to him that the presumed primordial rock had been molten after the strata had formed. He had read about angular unconformities as interpreted by Neptunists, and found an unconformity at Jedburgh where layers of greywacke in the lower layers of the cliff face have been tilted almost vertically before being eroded to form a level plane, under horizontal layers of Old Red Sandstone. In the spring of 1788 he took a boat trip along the Berwickshire coast with John Playfair and the geologist Sir James Hall, and found a dramatic unconformity showing the same sequence at Siccar Point. Playfair later recalled that «the mind seemed to grow giddy by looking so far into the abyss of time», and Hutton concluded a 1788 paper he presented at the Royal Society of Edinburgh, later rewritten as a book, with the phrase «we find no vestige of a beginning, no prospect of an end».

Both Playfair and Hall wrote their own books on the theory, and for decades robust debate continued between Hutton's supporters and the Neptunists. Georges Cuvier's paleontological work in the 1790s, which established the reality of extinction, explained this by local catastrophes, after which other fixed species repopulated the affected areas. In Britain, geologists adapted this idea into «diluvial theory» which proposed repeated worldwide annihilation and creation of new fixed species adapted to a changed environment, initially identifying the most recent catastrophe as the biblical flood.

19th Century

From 1830 to 1833 Charles Lyell's multi-volume *Principles of Geology* was published. The work's subtitle was "An attempt to explain the former changes of the Earth's surface by reference to causes now in operation". He drew his explanations from field studies conducted directly before he went to work on the founding geology text, and developed Hutton's idea that the earth was shaped entirely by slow-moving forces still in operation today, acting over a very long period of time. The terms *uniformitarianism* for this idea, and *catastrophism* for the opposing viewpoint, were coined by William Whewell in a review of Lyell's book. *Principles of Geology* was the most influential geological work in the middle of the 19th century.

Lyell's Uniformitarianism

According to Reijer Hooykaas (1963), Lyell's uniformitarianism is a family of four related propositions, not a single idea:

- Uniformity of law – the laws of nature are constant across time and space.
- Uniformity of methodology – the appropriate hypotheses for explaining the geological past are those with analogy today.
- Uniformity of kind – past and present causes are all of the same kind, have the same energy, and produce the same effects.
- Uniformity of degree – geological circumstances have remained the same over time.

None of these connotations requires another, and they are not all equally inferred by uniformitarians.

Gould explained Lyell's propositions in *Time's Arrow, Time's Cycle* (1987), stating that Lyell conflated two different types of propositions: a pair of methodological assumptions with a pair of substantive hypotheses. The four together make up Lyell's uniformitarianism.

Methodological Assumptions

The two methodological assumptions below are accepted to be true by the majority of scientists and geologists. Gould claims that these philosophical propositions must be assumed before you can proceed as a scientist doing science. "You cannot go to a rocky outcrop and observe either the constancy of nature's laws or the working of unknown processes. It works the other way around." You first assume these propositions and "then you go to the outcrop."

- **Uniformity of law across time and space**: Natural laws are constant across space and time.

The axiom of uniformity of law is necessary in order for scientists to extrapolate (by inductive inference) into the unobservable past.The constancy of natural laws must be assumed in the study of the past; else we cannot meaningfully study it.

- **Uniformity of process across time and space**: Natural processes are constant across time and space.

Though similar to uniformity of law, this second *a priori* assumption, shared by the vast majority of scientists, deals with geological causes, not physico-chemical laws. The past is to be explained by processes acting currently in time and space rather than inventing extra esoteric or unknown processes *without good reason*, otherwise known as parsimony or Occam's razor.

Substantive Hypotheses

The substantive hypotheses were controversial and, in some cases, accepted by few. These hypotheses are judged true or false on empirical grounds through scientific observation and repeated experimental data. This is in contrast with the previous two philosophical assumptions that come before one can do science and so cannot be tested or falsified by science.

- **Uniformity of rate across time and space**: Change is typically slow, steady, and gradual.

Uniformity of rate (or gradualism) is what most people (including geologists) think of when they hear the word "uniformitarianism," confusing this hypothesis with the entire definition. As late as 1990, Lemon, in his textbook of stratigraphy, affirmed that "The uniformitarian view of earth history held that all geologic processes proceed continuously and at a very slow pace."

Gould explained Hutton's view of uniformity of rate; mountain ranges or grand canyons are built by accumulation of nearly insensible changes added up through vast time. Some major events such as floods, earthquakes, and eruptions, do occur. But these catastrophes are strictly local. They neither occurred in the past, nor shall happen in the future, at any greater frequency or extent than they display at present. In particular, the whole earth is never convulsed at once.

- **Uniformity of state across time and space**: Change is evenly distributed throughout space and time.

The uniformity of state hypothesis implies that throughout the history of our earth there is no progress in any inexorable direction. The planet has almost always looked and behaved as it does now. Change is continuous, but leads nowhere. The earth is in balance: a dynamic steady state.

20th Century

Stephen Jay Gould's first scientific paper, *Is uniformitarianism necessary?* (1965), reduced these four interpretations to two. He dismissed the first principle, which asserted spatial and temporal invariance of natural laws, as no longer an issue of debate. He rejected the third (uniformity of rate) as an unjustified limitation on scientific inquiry, as it constrains past geologic rates and conditions to those of the present. So, Lyellian uniformitarianism was unnecessary.

Modern Uniformitarianism Includes Periodic Catastrophes

Uniformitarianism was originally proposed in contrast to catastrophism, which states that the distant past "consisted of epochs of paroxysmal and catastrophic action interposed between periods of comparative tranquility" Especially in the late 19th and early 20th centuries, most geologists took this interpretation to mean that catastrophic events are not important in geologic time; one example of this is the debate of the formation of the Channeled Scablands due to the catastrophic Missoula glacial outburst floods. An important result of this debate and others was the re-clarification that, while the same principles operate in geologic time, catastrophic events that are infrequent on human time-scales can have important consequences in geologic history. Derek Ager has noted that "geologists do not deny uniformitarianism in its true sense, that is to say, of interpreting the past by means of the processes that are seen going on at the present day, so long as we remember that the periodic catastrophe is one of those processes. Those periodic catastrophes make more showing in the stratigraphical record than we have hitherto assumed."

Even Charles Lyell thought that ordinary geological processes would cause Niagara Falls to move upstream to Lake Erie within 10,000 years, leading to catastrophic flooding of a large part of North America.

Modern geologists do not apply uniformitarianism in the same way as Lyell. They question if rates of processes were uniform through time and only those values measured during the history of geology are to be accepted. The present may not be a long enough key to penetrate the deep lock of the past. Geologic processes may have been active at different rates in the past that humans have not observed. «By force of popularity, uniformity of rate has persisted to our present day. For more than a century, Lyell's rhetoric conflating axiom with hypotheses has descended in unmodified form. Many geologists have been stifled by the belief that proper methodology includes an a priori commitment to gradual change, and by a preference for explaining large-scale phenomena as the concatenation of innumerable tiny changes.»

The current consensus is that Earth's history is a slow, gradual process punctuated by occasional natural catastrophic events that have affected Earth and its inhabitants. In practice it is reduced from Lyell's conflation to simply the two philosophical assumptions. This is also known as the principle of geological actualism, which states that all past geological action was like all present geological action. The principle of actualism is the cornerstone of paleoecology.

PRINCIPLE OF ORIGINAL HORIZONTALITY

The Principle of Original Horizontality states that layers of sediment are originally deposited horizontally under the action of gravity . It is a relative dating technique. The principle is important to the analysis of folded and tilted strata. It was first proposed by the Danish geological pioneer Nicholas Steno (1638–1686).

From these observations is derived the conclusion that the Earth has not been static and that great forces have been at work over long periods of time, further leading to the conclusions of the science of plate tectonics; that movement and collisions of large plates of the Earth's crust is the cause of folded strata.

As one of Steno's Laws, the Principle of Original Horizontality served well in the nascent days of geological science. However, it is now known that not all sedimentary layers are deposited purely horizontally. For instance, coarser grained sediments such as sand may be deposited at angles of up to 15 degrees, held up by the internal friction between grains which prevents them slumping to a lower angle without additional reworking or effort. This is known as the angle of repose, and a prime example is the surface of sand dunes.

A stratigraphic section of Ordovician rock exposed in central Tennessee, USA. The sediments composing these rocks were formed in an ocean and deposited in horizontal layers.

The Permian through Jurassic stratigraphy of the Colorado Plateau area of southeastern Utah is a great example of Original Horizontality. These strata make up much of the famous prominent rock formations in widely spaced protected areas such as Capitol Reef National Park and Canyonlands National Park. From top to bottom: Rounded tan domes of the Navajo Sandstone, layered red Kayenta Formation, cliff-forming, vertically jointed, red Wingate Sandstone, slope-forming, purplish Chinle Formation, layered, lighter-red Moenkopi Formation, and white, layered Cutler Formation sandstone. Picture from Glen Canyon National Recreation Area, Utah.

Similarly, sediments may drape over a pre-existing inclined surface: these sediments are usually deposited conformably to the pre-existing surface. Also sedimentary beds may pinch out along strike, implying that slight angles existed during their deposition. Thus the Principle of Original Horizontality is widely, but not universally, applicable in the study of sedimentology, stratigraphy and structural geology.

LAW OF SUPERPOSITION

Layer upon layer of rocks on north shore of Isfjord, Svalbard, Norway. Since there is no overturning, the rock at the bottom is older than the rock on the top by the Law of Superposition.

The **law of superposition** is an axiom that forms one of the bases of the sciences of geology, archaeology, and other fields dealing with geological stratigraphy. In its plainest form, it states that in undeformed stratigraphic sequences, the oldest strata will be at the bottom of the sequence. This is important to stratigraphic dating, which assumes that the law of superposition holds true and that an object cannot be older than the materials of which it is composed. The law was first proposed in the 17th century by the Danish scientist Nicolas Steno.

Archaeological Considerations

Superposition in archaeology and especially in stratification use during excavation is slightly different as the processes involved in laying down archaeological strata are somewhat different from geological processes. Man made intrusions and activity in the archaeological record need not form chronologically from top to bottom or be deformed from the horizontal as natural strata are by equivalent processes. Some archaeological strata (often termed as contexts or layers) are created by undercutting previous strata. An example would be that the silt back-fill of an underground drain would form some time after the ground immediately above it. Other examples of non vertical superposition would be modifications to standing structures such as the creation of new doors and windows in a wall. Superposition in archaeology requires a degree of interpretation to correctly identify chronological sequences and in this sense superposition in archaeology is more dynamic and multi-dimensional.

PRINCIPLE OF FAUNAL SUCCESSION

The **principle of faunal succession**, also known as the **law of faunal succession**, is based on the observation that sedimentary rock strata contain fossilized flora and fauna, and that these fossils succeed each other vertically in a specific, reliable order that can be identified over wide horizontal distances. A fossilized Neanderthal bone will never be found in the same stratum as a fossilized Megalosaurus, for example, because neanderthals and megalosaurs lived during different geological periods, separated by many millions of years. This allows for strata to be identified and dated by the fossils found within.

This principle, which received its name from the English geologist William Smith, is of great importance in determining the relative age of rocks and strata. The fossil content of rocks together with the law of superposition helps to determine the time sequence in which sedimentary rocks were laid down.

Evolution explains the observed faunal and floral succession preserved in rocks. Faunal succession was documented by Smith in England during the first decade of the 19th century, and concurrently in France by Cuvier (with the assistance of the mineralogist Alexandre Brongniart). Archaic biological features and organisms are succeeded in the fossil record by more modern versions. For instance, paleontologists investigating the evolution of birds predicted that feathers would first be seen in primitive forms on flightless predecessor organisms such as feathered dinosaurs. This is precisely what has been discovered in the fossil record: simple feathers, incapable of supporting flight, are succeeded by increasingly large and complex feathers.

In practice, the most useful **diagnostic species** are those with the fastest rate of species turnover and the widest distribution; their study is termed biostratigraphy, the science of dating rocks by using the fossils contained within them. In Cenozoic strata, fossilized tests of foraminifera are often used to determine faunal succession on a refined scale, each biostratigraphic unit (biozone) being a geological stratum that is defined on the basis of its characteristic fossil taxa. An outline microfaunal zonal scheme based on both foraminifera and ostracoda was compiled by M. B. Hart (1972).

Simply, the earlier fossil life forms are simpler than more recent forms, and more recent forms are most similar to existing forms (principle of faunal succession).

REFERENCES

1. Cross Cutting. World of Earth Science. Ed. K. Lee Lerner and Brenda Wilmoth Lerner. Gale Cengage, 2003.

2. Nicolai Stenonis solido intra solidum naturaliter contento dissertationis prodromus ... Florentiae : ex typographia sub signo Stellae (1669)
3. Hutton, James. Theory of the Earth, 1795
4. Lyell, Charles. Principles of Geology, 1830

CHAPTER 4

Geologic Province

Chapter Contents

- Introduction
- Geologic Provinces by Resources
- Shield (Geology)
- Platform (Geology)
- Orogeny
- History Of The Concept
- Structural Basin
- Large Igneous Province

The geologic province and thermo-tectonic age maps were compiled in 1997 to ensure a consistent method of providing values for these two parameters. The maps depict only features approximately 150 km across and greater due to the fact that the resolution of the maps should be consistent with the resolution of the seismic refraction data.

INTRODUCTION

Geologic provinces of the World (USGS)

Shield	Oceanic crust:
Platform	0–20 Ma
Orogen	20–65 Ma
Basin	>65 Ma
Large igneous province	
Extended crust	

A **geologic or geomorphic province** is a spatial entity with common geologic or geomorphic attributes. A province may include a single dominant structural element such as a basin or a fold belt, or a number of contiguous related elements. Adjoining provinces may be similar in structure but be considered separate due to differing histories.

Geologic provinces by origin

Province	**Definition**	**Subcategories**	**Examples**
Shield	Exposed Precambrian crystalline igneous and metamorphic rocks that form tectonically stable areas		• Arabian-Nubian Shield • Canadian Shield
Platform	Horizontal or gently-lying sedimentary strata covering a basement of igneous or metamorphic rocks	• Carbonate platform	
Orogen	Linear or arc-shaped formation where continental crust has been folded, deformed and uplifted to form mountain ranges	• Island arc • Continental arc • Forearc	
Basin	Low-lying formation of rock strata formed by tectonic warping of previously horizontal strata	• Cratonic basin • Foredeep basin • Sedimentary basin	

Large igneous province	Accumulation of igneous rocks, including liquid rock (intrusive) or volcanic rock formations (extrusive)	• North Atlantic Igneous Province • Columbia River Basalt Group	
Extended crust	Continental crust thinned due to extensional strain	• Passive margin • Rift	• Basin and Range Province • Northern Cordilleran Volcanic Province

GEOLOGIC PROVINCES BY RESOURCES

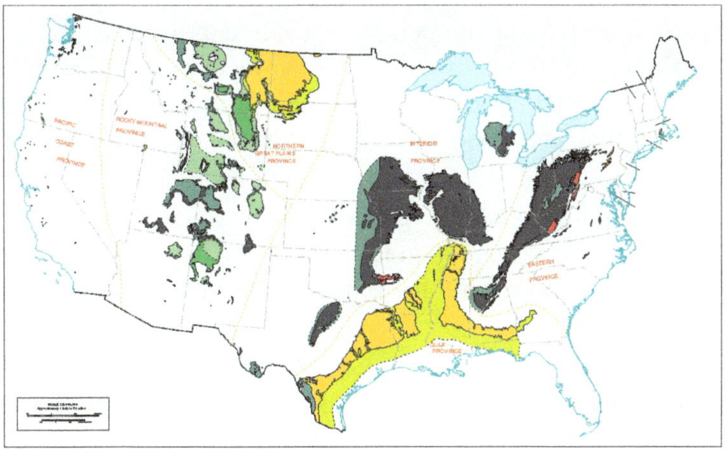

US coal regions and Provinces.

Some studies classify provinces based upon mineral resources, such as mineral deposits. There are a particularly large number of provinces identified worldwide for petroleum and other mineral fuels, such as the Niger Delta petroleum province.

SHIELD (GEOLOGY)

Geologic provinces of the World (USGS)

A **shield** is generally a large area of exposed Precambrian crystalline igneous and high-grade metamorphic rocks that form tectonically stable areas. In all cases, the age of these rocks is greater than 570 million years and sometimes dates back 2 to 3.5 billion years.

They have been little affected by tectonic events following the end of the Precambrian, and are relatively flat regions where mountain building, faulting, and other tectonic processes are greatly diminished compared with the activity that occurs at the margins of the shields and the boundaries between tectonic plates.

The term *shield*, used to describe this type of geographic region, appears in the 1901 English translation of *Eduard Suess's Face of Earth* by H. B. C. Sollas, and comes from the shape "not unlike a flat shield" of the Canadian Shield which has an outline that "suggests the shape of the shields carried by soldiers in the days of hand-to-hand combat."

A shield is that part of the continental crust in which these usually Precambrianbasement rocks crop out extensively at the surface. Shields themselves can be very complex: they consist of vast areas of granitic or granodioriticgneisses, usually of tonalitic composition, and they also contain belts of sedimentary rocks, often surrounded by low-grade volcano-sedimentary sequences, or greenstone belts. These rocks are frequently metamorphosed greenschist, amphibolite, and granulite facies.

Shields are normally the nucleus of continents and most are bordered by belts of folded Cambrian rocks. Because of their stability, erosion has flattened the topography of most of the continental shields; however, they commonly do have a very gently convex surface. They are also surrounded by a sediment-covered platform. By contrast, in a platform, the shield (more accurately referred to then as the crystalline basement) is overlain by horizontal or subhorizontal sediment. Together, the shield, platform and basement are the parts that comprise the stable interior portion of the continental crust known as the craton.

The margins surrounding a shield generally constitute relatively mobile zones of intense tectonic or plate-like dynamic mechanisms. In these areas, complex sequences of mountain building (orogeny) events have been documented over the past few hundred million years.

For example, the Ural Mountains, to the west of the Angaran Shield, are atop the mobile zone that separates the shield from the Baltic Shield. Similarly, the Himalayas are on the mobile boundary between the Angaran and Indian shields. Shield margins have been subject to geotectonic forces that have both destroyed and rebuilt the margins and the cratons that they partially comprise. In fact, the growth of continents has occurred as a result of the accretion of younger rocks that underwent deformations during series of mountain building processes. In a sense, these belts of folded rocks have been welded onto the borders of the preexisting shields, thus increasing the size of the proto-continents that they make up.

Continental shields occur on all continents, for example:

- The Canadian Shield forms the nucleus of North America and extends from Lake Superior on the south to the Arctic Islands on the north, and from western Canada eastward across to include most of Greenland.
- The Amazonian (Brazilian) Shield on the eastern bulge portion of South America. Bordering this is the Guiana Shield to the north, and the Platian Shield to the south.
- The Baltic (Fennoscandian) Shield is located in eastern Norway, Finland and

Sweden.

- The African (Ethiopian) Shield is located in Africa.
- The Australian Shield occupies most of the western half of Australia.
- The Arabian-Nubian Shield on the western edge of Arabia.
- The Antarctic Shield.
- In Asia, an area in China and North Korea is sometimes referred to as the China-Korean Shield.
- The Angaran Shield, as it is sometimes called, is bounded by the Yenisey River on the west, the Lena River on the east, the Arctic Ocean on the north, and Lake Baikal on the south.
- The Indian Shield occupies two-thirds of the southern Indian peninsula.

PLATFORM (GEOLOGY)

In geology, a **platform** is a continental area covered by relatively flat or gently tilted, mainly sedimentary strata, which overlie a basement of consolidated igneous or metamorphic rocks of an earlier deformation. Platforms, shields and the basement rocks together constitute cratons. Platform sediments can be classified into the following groups: a protoplatform of metamorphosed sediments at the bottom, a "quasiplatform" of slightly deformed sediments, a "cataplatform", and a "orthoplatform" at the top. The Mesoproterozoic Jotnian sediments of the Baltic area are examples of a "quasiplatform". The post-Ordovician rocks of the South American Platform are examples of an orthoplatform.

OROGENY

Orogeny refers to forces and events leading to a large structural deformation of the Earth's lithosphere (crust and uppermost mantle) due to the interaction between tectonic plates. Orogens or orogenic belts develop when a continental plate is crumpled and is pushed upwards to form mountain ranges, and involve a great range of geological processes collectively called orogenesis.

Orogeny is the primary mechanism by which mountains are built on continents. The word orogeny comes from Ancient Greek (ρος *oros*, "mountain" + γένεσις *genesis* for "creation, origin"). Though it was used before him, the term was employed by the American geologist G.K. Gilbert in 1890 to describe the process of mountain building as distinguished from epeirogeny.

Physiography

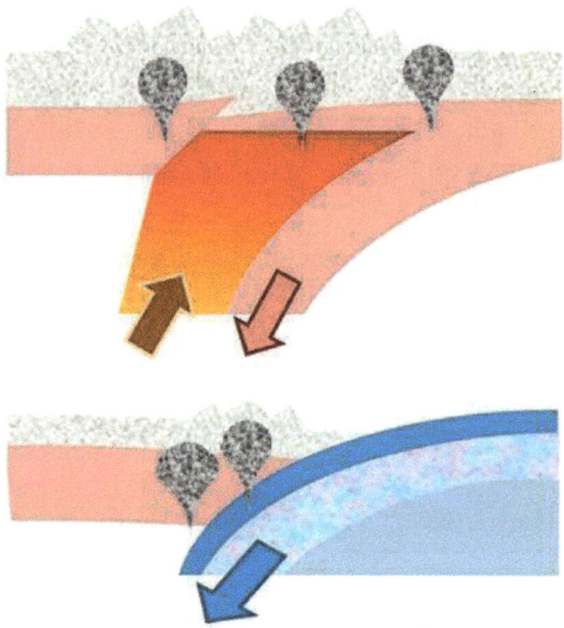

Two processes that can contribute to an orogen. Top: delamination by intrusion of hot asthenosphere; Bottom: Subduction of ocean crust. The two processes lead to differently located granites (bubbles in diagram), providing evidence as to which process actually occurred.

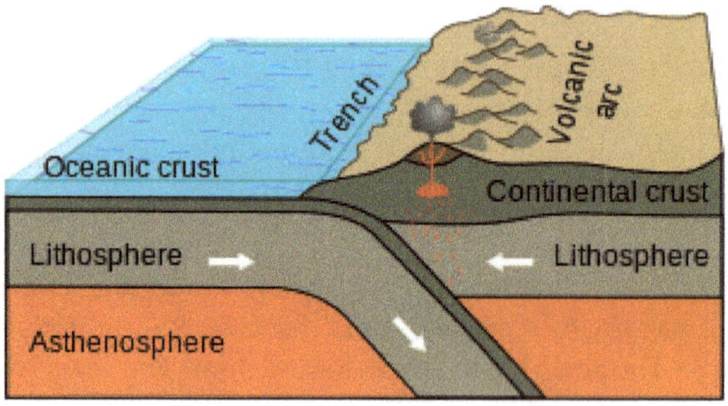

Subduction of an oceanic plate by a continental plate to form a noncollisional orogen. (example: the Andes)

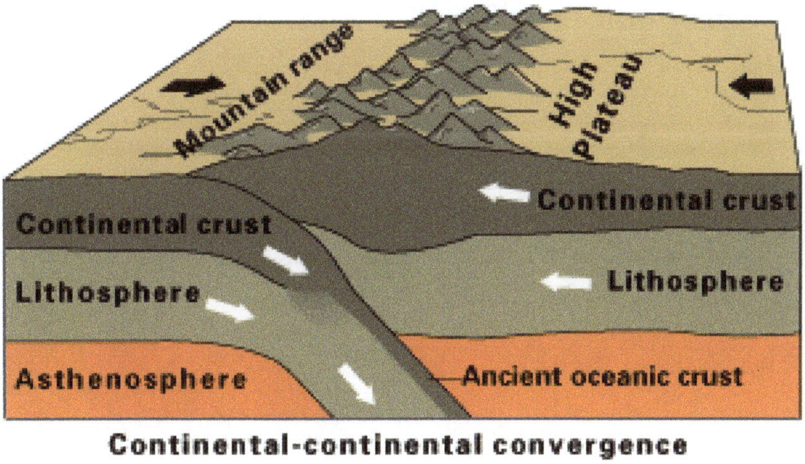

Continental collision of two continental plates to form a collisional orogen. However, usually no continental crust is subducted, only uplifted. (example: the Alps)

Formation of an orogen is accomplished in part by the tectonic processes of subduction (where a continent rides forcefully over an oceanic plate (noncollisional orogens)) or convergence of two or more continents (collisional orogens).

Orogeny usually produces long arcuate (from Latin *arcuare*, "to bend like a bow") structures, known as *orogenic belts*. Generally, orogenic belts consist of long parallel strips of rock exhibiting similar characteristics along the length of the belt. Orogenic belts are associated with subduction zones, which consume crust, produce volcanoes, and build island arcs. Geologists attribute the arcuate structure to the rigidity of the descending plate, and island arc cusps relate to tears in the descending lithosphere. These island arcs may be added to a continent during an orogenic event.

The processes of orogeny can take tens of millions of years and build mountains from plains or from the seabed. The topographic height of orogenic mountains is related to the principle of isostasy, that is, a balance of the downward gravitational force upon an upthrust mountain range (composed of light, continental crust material) and the buoyant upward forces exerted by the dense underlying mantle.

Frequently, rock formations that undergo orogeny are severely deformed and undergo metamorphism. Orogenic processes may push deeply buried rocks to the surface. Sea-bottom and near-shore material may cover some or all of the orogenic area. If the orogeny is due to two continents colliding, very high mountains can result (see Himalayas).

An orogenic event may be studied: (a) as a tectonic structural event, (b) as a geographical event, and (c) as a chronological event.

Orogenic events:

- cause distinctive structural phenomena related to tectonic activity
- affect rocks and crust in particular regions, and
- happen within a specific period

Orogen (or "orogenic system")

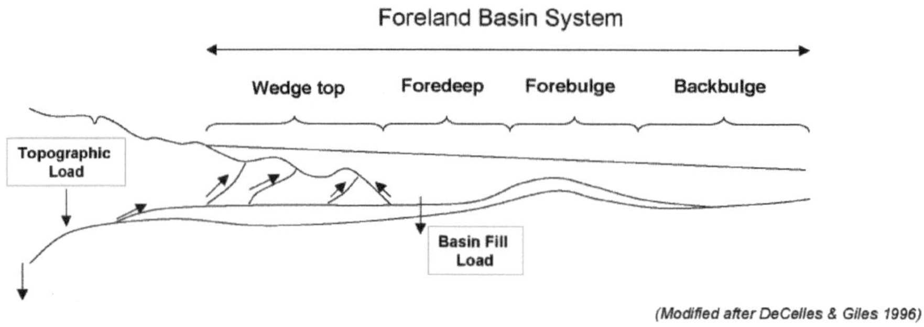

The Foreland Basin System

An orogeny produces an *orogen*, or (mountain) range-*foreland basin* system.

The *foreland basin* forms ahead of the orogen due mainly to loading and resulting flexure of the lithosphere by the developing mountain belt. A typical foreland basin is subdivided into a wedge-top basin above the active orogenic wedge, the foredeep immediately beyond the active front, a forebulge high of flexural origin and a back-bulge area beyond, although not all of these are present in all foreland-basin systems. The basin migrates with the orogenic front and early deposited foreland basin sediments become progressively involved in folding and thrusting. Sediments deposited in the foreland basin are mainly derived from the erosion of the actively uplifting rocks of the mountain range, although some sediments derive from the foreland. The fill of many such basins shows a change in time from deepwater marine (*flysch*-style) through shallow water to continental (*molasse*-style) sediments.

Orogenic Cycle

Although orogeny involves plate tectonics, the tectonic forces result in a variety of associated phenomena, including magmatism, metamorphism, crustal melting, and crustal thickening. What exactly happens in a specific orogen depends upon the strength and rheology of the continental lithosphere, and how these properties change during orogenesis.

In addition to orogeny, the orogen (once formed) is subject to other processes, such as sedimentation and erosion. The sequence of repeated cycles of sedimentation, deposition and erosion, followed by burial and metamorphism, and then by formation of granitic batholiths and tectonic uplift to form mountain chains, is called the *orogenic cycle*. For example, the *Caledonian Orogeny* refers to the Silurian and Devonian events that resulted from the collision of Laurentia with Eastern Avalonia and other former fragments of Gondwana. The *Caledonian Orogen* resulted from these events and various others that are part of its peculiar orogenic cycle.

In summary, an orogeny is a long-lived deformational episode during which many geological phenomena play a role. The orogeny of an orogen is only part of the orogen›s orogenic cycle.

Erosion

Erosion represents a subsequent phase of the orogenic cycle. Erosion inevitably removes much of the mountains, exposing the core or *mountain roots* (metamorphic rocks brought to the surface from a depth of several kilometres). Isostatic movements may help such exhumation by balancing out the buoyancy of the evolving orogen. Scholars debate about the extent to which erosion modifies the patterns of tectonic deformation (see erosion and tectonics). Thus, the final form of the majority of old orogenic belts is a long arcuate strip of crystalline metamorphic rocks sequentially below younger sediments which are thrust atop them and which dip away from the orogenic core.

An orogen may be almost completely eroded away, and only recognizable by studying (old) rocks that bear traces of orogenesis. Orogens are usually long, thin, arcuate tracts of rock that have a pronounced linear structure resulting in terranes or blocks of deformed rocks, separated generally by suture zones or dipping thrust faults. These thrust faults carry relatively thin slices of rock (which are called nappes or thrust sheets, and differ from tectonic plates) from the core of the shortening orogen out toward the margins, and are intimately associated with folds and the development of metamorphism.

Biology

In the 1950s and 1960s the study of orogeny, coupled with biogeography (the study of the distribution and evolution of flora and fauna),geography and mid ocean ridges, contributed greatly to the theory of plate tectonics. Even at a very early stage, life played a significant role in the continued existence of oceans, by affecting the composition of the atmosphere. The existence of oceans is critical to sea-floor spreading and subduction.

Relationship to Mountain Building

An example of thin-skinned deformation (thrust faulting) of the Sevier Orogeny in Montana. Note the white Madison Limestone repeated, with one example in the foreground (that pinches out with distance) and another to the upper right corner and top of the picture.

Sierra Nevada Mountains (a result of delamination) as seen from the International Space Station.

Mountain formation occurs through a number of mechanisms.

Mountain complexes result from irregular successions of tectonic responses due to sea-floor spreading, shifting lithosphere plates, transform faults, and colliding, coupled and uncoupled continental margins.

—Peter J Coney

Large modern orogenies often lie on the margins of present-day continents; the Alleghenian (Appalachian), Laramide, and Andean orogenies exemplify this in the Americas. Older inactive orogenies, such as the Algoman, Penokean and Antler, are represented by deformed rocks and sedimentary basins further inland.

Areas that are rifting apart, such as mid-ocean ridges and the East African Rift, have mountains due to thermal buoyancy related to the hot mantle underneath them; this thermal buoyancy is known as dynamic topography. In strike-slip orogens, such as the San Andreas Fault, restraining bends result in regions of localized crustal shortening and mountain building without a plate-margin-wide orogeny. Hotspot volcanism results in the formation of isolated mountains and mountain chains that are not necessarily on tectonic-plate boundaries.

Regions can also experience uplift as a result of delamination of the lithosphere, in which an unstable portion of cold lithospheric root drips down into the mantle, decreasing the density of the lithosphere and causing buoyant uplift. An example is the Sierra Nevada in California. This range of fault-block mountains experienced renewed uplift after a delamination of the lithosphere beneath them.

Finally, uplift and erosion related to epeirogenesis (large-scale vertical motions of portions of continents without much associated folding, metamorphism, or deformation) can create local topographic highs.

Mount Rundle, Banff, Alberta.

Mount Rundle on the TransCanada Highway between Banff and Canmore provides a classic example of a mountain cut in dipping-layered rocks. Millions of years ago a collision caused an orogeny forcing horizontal layers of an ancient ocean crust to be thrust up at an angle of 50–60°. That left Rundle with one sweeping, tree-lined smooth face, and one sharp, steep face where the edge of the uplifted layers are exposed.

The collision causing the Columbia Orogeny occurred about 175 million years ago, and as the shock wave moved eastward, it forced huge masses of rock to crack and slide up over its neighbours. This is known as thrust faulting and was instrumental in the formation of the Rockies. The shock wave began piling up the western ranges, and then the main ranges, around 120 million years ago.

—Mountains in Nature

HISTORY OF THE CONCEPT

Before the development of geologic concepts during the 19th century, the presence of marine fossils in mountains was explained in Christian contexts as a result of the Biblical Deluge. This was an extension of Neoplatonic thought, which influenced early Christian writers.

The 13th-century Dominican scholar Albert the Great posited that, as erosion was known to occur, there must be some process whereby new mountains and other land-forms were thrust up, or else there would eventually be no land; he suggested that marine fossils in mountainsides must once have been at the sea-floor. Orogeny was used by Amanz Gressly (1840) and Jules Thurmann (1854) as *orogenic* in terms of the creation of mountain elevations, as the term *mountain building* was still used to describe the processes. Elie de Beaumont (1852) used the evocative "Jaws of a Vise" theory to explain orogeny, but was more concerned with the height rather than the implicit structures created by and contained in orogenic belts. His theory essentially held that mountains were created by the squeezing of certain rocks. Eduard Suess (1875) recognised the importance of horizontal movement of rocks. The concept of a *precursor geosyncline* or initial downward warping of the solid earth (Hall, 1859) prompted James Dwight Dana (1873) to include the concept of *compression* in the theories surrounding mountain-building. With hindsight, we can discount Dana's conjecture that this contraction was due to the cooling of the Earth (aka the cooling Earth theory). The cooling Earth theory was the chief paradigm for most geologists until the 1960s. It was, in the context of orogeny, fiercely contested by proponents of vertical movements in the crust (similar to tephrotectonics), or convection within the asthenosphere or mantle.

Gustav Steinmann (1906) recognised different classes of orogenic belts, including the *Alpine type orogenic belt*, typified by a flysch and molasse geometry to the sediments; ophiolite sequences, tholeiitic basalts, and a nappe style fold structure.

In terms of recognising orogeny as an *event*, Leopold von Buch (1855) recognised that orogenies could be placed in time by bracketing between the youngest deformed rock and the oldest undeformed rock, a principle which is still in use today, though commonly investigated by geochronology using radiometric dating.

H.J. Zwart (1967) drew attention to the metamorphic differences in orogenic belts, proposing three types, modified by W. S. Pitcher in 1979and further modified as:

- Hercynotype (back-arc basin type);
    - Shallow, low-pressure metamorphism; thin metamorphic zones
    - Metamorphism dependent on increase in temperature
    - Abundant granite and migmatite
    - Few ophiolites, ultramafic rocks virtually absent
    - very wide orogen with small and slow uplift
    - nappe structures rare
- Alpinotype (ocean trench style);
    - deep, high pressure, thick metamorphic zones
    - metamorphism of many facies, dependent on decrease in pressure
    - few granites or migmatites
    - abundant ophiolites with ultramafic rocks
    - Relatively narrow orogen with large and rapid uplift
    - Nappe structures predominant

- Cordilleran (arc) type;
    - dominated by calc-alkaline igneous rocks, andesites, granite batholiths
    - general lack of migmatites, low geothermal gradient
    - lack of ophiolite and abyssal sedimentary rocks (black shale, chert, etcetera)
    - low-pressure metamorphism, moderate uplift
    - lack of nappes

The advent of plate tectonics has explained the vast majority of orogenic belts and their features. The cooling earth theory (principally advanced by Descartes) is dispensed with, and tephrotectonic style vertical movements have been explained primarily by the process of isostasy.

Some oddities exist, where simple collisional tectonics are modified in a transform plate boundary, such as in New Zealand, or where island arc orogenies, for instance in New Guinea occur away from a continental backstop. Further complications such as Proterozoic continent-continent collisional orogens, explicitly the Musgrave Block in Australia, previously inexplicable are being brought to light with the advent of seismic imaging techniques which can resolve the deep crust structure of orogenic belts.

STRUCTURAL BASIN

A **structural basin** is a large-scale structural formation of rock strata formed by tectonic warping of previously flat-lying strata. Structural basins are geological depressions, and are the inverse of domes. Some elongated structural basins are also known as synclines. Structural basins may also be sedimentary basins, which are aggregations of sediment that filled up a depression or accumulated in an area; however, many structural basins were formed by tectonic events long after the sedimentary layers were deposited.

Basins appear on a geologic map as roughly circular or elliptical, with concentric layers. Because the strata dip toward the center, the exposed strata in a basin are progressively younger from the outside in, with the youngest rocks in the center. Basins are often large in areal extent, often hundreds of kilometers across.

Structural basins are often important sources of coal, petroleum, and groundwater.

Examples of Structural Basins

Europe

- Hampshire Basin, United Kingdom
- London Basin, United Kingdom
- Paris Basin, France

- Permian Basin, Poland, northern Germany, Denmark, the Netherlands, the North Sea, and Scotland

North America

Trinidad and Tobago

- Southern Basin, Trinidad

United States

- Albuquerque Basin, New Mexico
- Appalachian Basin, Eastern United States
- Big Horn Basin, Wyoming
- Black Warrior Basin, Alabama and Mississippi
- Delaware Basin, Texas and New Mexico
- Denver Basin, Colorado
- Illinois Basin, Illinois
- Los Angeles Basin, California
- Michigan Basin, Michigan
- North Park Colorado Basin
- Paradox Basin, Utah and Colorado
- Permian Basin, Texas and New Mexico
- Piceance Basin, Colorado
- Powder River Basin, Wyoming and Montana
- Raton Basin, Colorado and New Mexico
- Sacramento Basin, California
- San Juan Basin, New Mexico and Colorado
- Uinta Basin, Utah
- Williston Basin, Montana and North Dakota
- Wind River Basin, Wyoming

Oceania

Australia

- Amadeus Basin
- Bowen Basin
- Cooper Basin
- Galilee Basin

South America

- Chaco Basin, Argentina, Bolivia and Paraguay
- Magallanes Basin, Chile
- Neuquén Basin, Argentina and Chile
- Paraná Basin, Argentina, Brazil, Paraguay and Uruguay

LARGE IGNEOUS PROVINCE

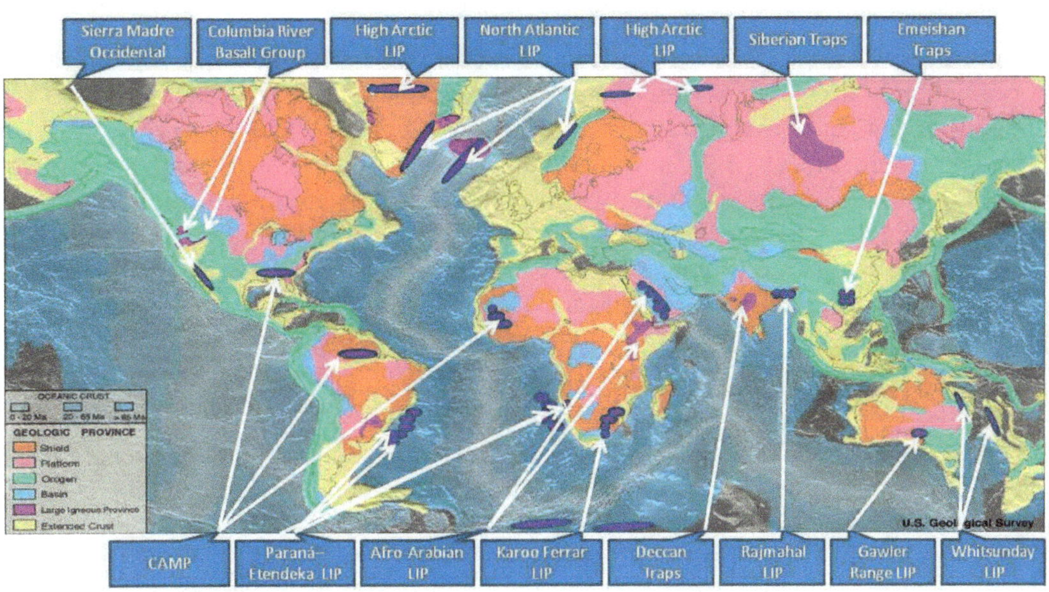

Only a few of the largest large igneous provinces appear (coloured dark purple) on this geological map, which depicts crustal geologic provinces as seen in seismic refraction data.

In geology, a large igneous province (LIP) is an extremely large accumulation of igneous rocks, including liquid rock (intrusive) or volcanic rock formations (extrusive), arising when hot magma extrudes from inside the Earth and flows out. The source of many or all LIPs is variously attributed to mantle plumes or to processes associated with plate tectonics. Types of LIPs can include *large volcanic provinces* (LVP), created through flood basalt, and *large plutonic provinces* (LPP). Rampino and Stothers (1988) cite eleven distinct flood-basalt episodes - occurring in the past 250 million years - which created volcanic provinces and plateaus and coincided with mass extinctions. However, more flood basalts have been recognized, such as the large Ontong Java Plateau, and the medium-sized Chilcotin Group, though the latter may be linked to the Columbia River Basalt Group. Formation depends on a range of factors, such as continental configuration, latitude, volume, rate, duration of eruption, style and setting (continental vs. oceanic), the pre-existing climate state, and the biota resilience to change.

Definition

In 1992 researchers first used the term *large igneous province* to describe very large accumulations—areas greater than 100,000 square kilometers (approximately the area of Iceland)—of mafic igneous rocks that were erupted or emplaced at depth within an extremely short geological time interval: a few million years or less. Mafic, basalt sea floors and other geological products of 'normal' plate tectonics were not included in the definition.

Types

The definition of *LIP* has been expanded and refined, and is still a work in progress. *LIP* is now frequently also used to describe voluminous areas of, not just mafic, but all types of igneous rocks. Sub-categorization of LIPs into *large volcanic provinces* (LVP) and *large plutonic provinces* (LPP), and including rocks produced by normal plate tectonic processes, has been proposed. Some LIPs are now intact, such the basaltic Deccan Traps in India, while others have been dismembered by plate tectonic motion, like the basaltic Central Atlantic Magmatic Province (CAMP)—parts of which are found in Brazil, eastern North America, and north-western Africa.

Motivations for study of LIPs

Large igneous provinces (LIPs) are created during short-lived igneous events resulting in relatively rapid and high-volume accumulations of volcanic and intrusive igneous rock. These events warrant study because:

- The study of LIPs has economic implications. Some workers associate them with trapped hydrocarbons. They are associated with economic concentrations of copper–nickel and iron. They are also associated with formation of major mineral provinces including Platinum-Group Element (PGE) Deposits, and in the Silicic LIPs, silver and gold deposits. Titanium and vanadium deposits are also found in association with LIPs.

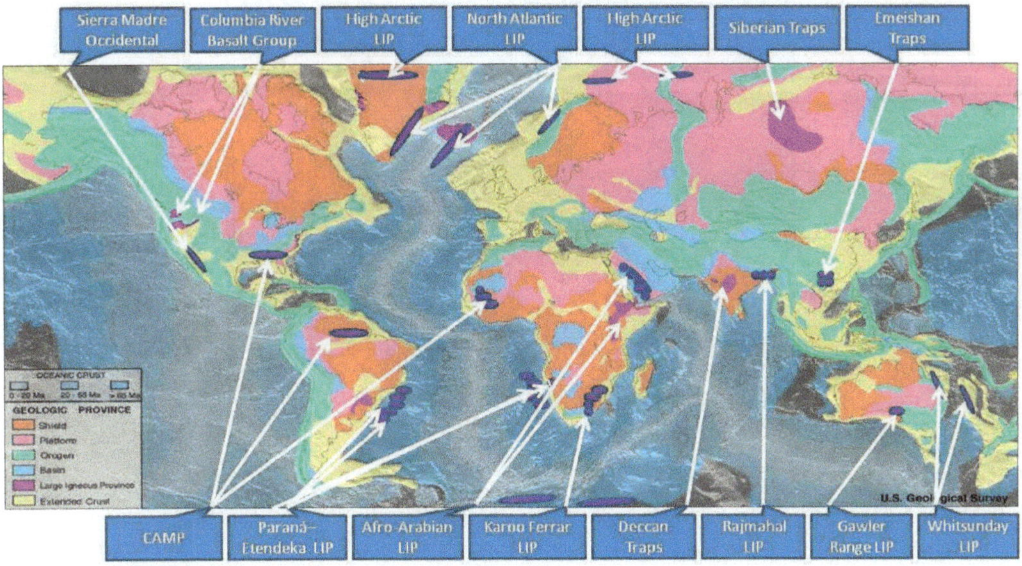

Map showing the recognized continental large igneous provinces.

- Future LIP events may pose a hazard to human civilization. LIPs in the geological record have marked major changes in the hydrosphere and atmosphere, leading to major climate shifts and maybe mass extinctions of species.
- Plate tectonic theory explains topography using interactions between the tectonic plates, as influenced by viscous stresses created by flow within the underlying mantle. Since the mantle is extremely viscous, the mantle flow rate varies in pulses which are reflected in the lithosphere by small amplitude, long wavelength undulations. Understanding how the interaction between mantle flow and lithosphere elevation influences formation of LIPs is important to gaining insights into past mantle dynamics.
- LIPs have played a major role in continental breakup, continental formation, new crustal additions from the upper mantle, and supercontinent cycles.

Large Igneous Province Formation

Earth has an outer shell made of a number of discrete, moving tectonic plates floating on a solid convective mantle above a liquid core. The mantle›s flow is driven by the descent of cold tectonic plates during subduction and the complementary ascent of plumes of hot material from lower levels. The surface of the earth reflects stretching, thickening and bending of the tectonic plates as they interact.

Three Devil's grade in Moses Coulee, Washington is part of the Columbia River Basalt Group LIP.

Ocean-plate creation at upwellings, spreading and subduction are well accepted fundamentals of plate tectonics, with the upwelling of hot mantle materials and the sinking of the cooler ocean plates driving the mantle convection. In this model, tectonic plates diverge at mid-ocean ridges, where hot mantle rock flows upward to fill the space. Plate-tectonic processes account for the vast majority of Earth›s volcanism.

Beyond the effects of convectively driven motion, deep processes have other influences on the surface topography. The convective circulation drives up-wellings and down-wellings in Earth›s mantle that are reflected in local surface levels. Hot mantle materials rising up in a plume can spread out radially beneath the tectonic plate causing regions of uplift. These ascending plumes play an important role in LIP formation.

Formation Characteristics

When created, LIPs often have an areal extent of a few million km^2 and volumes on the order of 1 million km^3. In most cases, the majority of a basaltic LIP's volume is emplaced in less than 1 million years. One of the conundra of such LIPs' origins is to understand how enormous volumes of basaltic magma are formed and erupted over such short time scales, with effusion rates up to an order of magnitude greater than mid-ocean ridge basalts.

Formation Theories

The source of many or all LIPs are variously attributed to mantle plumes, to processes associated with plate tectonics or to meteorite impacts.

Plume Formation of LIPs

Although most of volcanic activity on Earth is associated with subduction zones or mid-oceanic ridges, there are significant regions of long-lived, extensive volcanism, known as hot spots, which are only indirectly related to plate tectonics. The Emperor Sea Mount chain, located on the Pacific plate, is one example, tracing millions of years of relative motion as the plate moves over the Hawaiian hot spot. Numerous hot spots of varying size and age have been identified across the world. These hot spots move slowly with respect to one another, but move an order of magnitude more quickly with respect to tectonic plates, providing evidence that they are not directly linked to tectonic plates.

The origin of hotspots remains controversial. Hotspots that reach the Earth's surface may have three distinct origins. The deepest probably originate from the boundary between the lower mantle and the core; roughly 15–20% have characteristics such as presence of a linear chain of sea mounts with increasing ages, LIPs at the point of origin of the track, low shear wave velocity indicating high temperatures below the current location of the track, and ratios of He^3 to He^4 which are judged consistent with a deep origin. Others such as the Pitcairn, Samoan and Tahitian hotspots appear to originate at the top of large, transient, hot lava domes (termed superswells) in the mantle. The remainder appear to originate in the upper mantle and have been suggested to result from the breakup of subducting lithosphere.

Recent imaging of the region below known hot spots (e.g., Yellowstone and Hawaii) using seismic-wave tomography has produced mounting evidence that supports relatively narrow, deep-origin, convective plumes that are limited in region compared to the large-scale plate tectonic circulation in which they are imbedded. Images reveal continuous, but torturous vertical paths with varying quantities of elevated temperature material, particularly at elevations where crystallographic transformations are predicted to occur.

Plate-related Stress formation of LIPs

A major alternative to the plume model is a model in which ruptures are caused by plate-related stresses that fractured the lithosphere, allowing melt to reach the surface from shallow heterogeneous sources. The high volumes of molten material that form the LIPs is postulated to be caused by convection in the upper mantle, which is secondary to the convection driving tectonic plate motion.

Early Formed Reservoir Outpourings

It has been proposed that geochemical evidence supports an early-formed reservoir that survived in the Earth's mantle for about 4.5 billion years. Molten material is postulated to have originated from this reservoir, contributing the Baffin Island flood basalt ~60 million years ago. Basalts from the Ontong Java plateau show similar isotopic and trace element signatures proposed for the early-Earth reservoir.

Meteorite-induced Formation

Seven pairs of hotspots and LIPs located on opposite sides of the earth have been noted;

analyses indicate this coincident antipodal location is highly unlikely to be random. The hotspot pairs include a large igneous province with continental volcanism opposite an oceanic hotspot. Oceanic impacts of large meteorites are expected to have high efficiency in converting energy into seismic waves. These waves would propagate around the world and reconverge close to the antipodal position; small variations are expected as the seismic velocity varies depending upon the route characteristics along which the waves propagate. As the waves focus on the antipodal position, they put the crust at the focal point under significant stress and are proposed to rupture it, creating antipodal pairs. When the meteorite impacts a continent, the lower efficiency of kinetic energy conversion into seismic energy is not expected to create an antipodal hotspot.

A second impact-related model of hotspot and LIP formation has been suggested in which minor hotspot volcanism was generated at large-body impact sites and flood basalt volcanism was triggered antipodally by focused seismic energy. This model has been challenged because impacts are generally considered seismically too inefficient , and the Deccan Traps of India were not antipodal to, and began erupting several Myr before, the end-Cretaceous Chicxulub impact in Mexico. In addition, no clear example of impact-induced volcanism, unrelated to melt sheets, has been confirmed at any known terrestrial crater.

Classification

In 1992, Coffin and Eldholm initially defined the term "large igneous province" (LIP) as representing a variety of mafic igneous provinces with areal extent >1 x 10^5 km^2 that represented "massive crustal emplacements of predominantly mafic (Mg- and Fe-rich) extrusive and intrusive rock, and originated via processes other than 'normal' seafloor spreading." That original definition included continental flood basalts, oceanic plateaus, large dike swarms (the eroded roots of a volcanic province), and volcanic rifted margins. Most of these LIPs consist of basalt, but some contain large volumes of associated rhyolite (e.g. the Columbia River Basalt Group in the western United States); the rhyolite is typically very dry compared to island arc rhyolites, with much higher eruption temperatures (850 °C to 1000 °C) than normal rhyolites.

Since 1992 the definition of ‹LIP› has been expanded and refined, and remains a work in progress. Some new definitions of the term ‹LIP› include large granitic provinces such as those found in the Andes Mountains of South America and in western North America. Comprehensive taxonomies have been developed to focus technical discussions.

In 2008, Bryan and Ernst refined the definition to narrow it somewhat: Large Igneous Provinces are magmatic provinces with areal extents >1 x 10^5 km^2, igneous volumes >1 x 10^5 km^3 and maximum lifespans of ~50 Myr that have intraplate tectonic settings or geochemical affinities, and are characterised by igneous pulse(s) of short duration (~1–5 Myr), during which a large proportion (>75%) of the total igneous volume has been emplaced. They are dominantly mafic, but also can have significant ultramafic and silicic components, and some are dominated by silicic magmatism." This definition places emphasis on the high magma emplacement rate characteristics of the LIP event and excludes seamounts, seamount groups, submarine ridges and anomalous seafloor crust. 'LIP' is now frequently used to also describe voluminous areas of, not just mafic, but all types of igneous rocks. Sub-categorization of LIPs into Large Volcanic Provinces (LVP) and Large Plutonic Provinces (LPP), and including rocks produced by 'normal' plate tectonic processes, has been proposed. Further, the threshold to be included as a LIP has been lowered to >$5x10^4$ km^2. The working taxonomy, focused heavily on geochemistry, which will be used to structure examples below, is:

- Large igneous provinces (LIP)
    - Large volcanic provinces (LVP)
        - Large rhyolitic provinces (LRPs)
        - Large andesitic provinces (LAPs)
        - Large basaltic provinces (LBPs): oceanic, or continental flood basalts
        - Large basaltic–rhyolitic provinces (LBRPs)
    - Large plutonic provinces (LPP)
        - Large granitic provinces (LGP)
        - Large mafic plutonic provinces

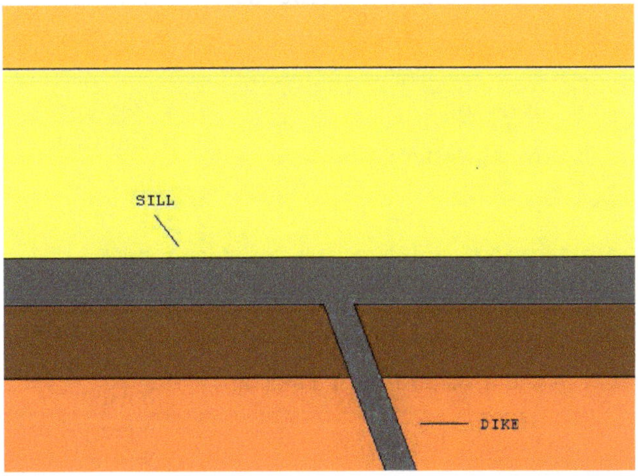

Illustration showing a vertical dike and a horizontal sill.

Aerally extensive dike swarms, sill provinces, and large layered ultramafic intrusions are indicators of LIPs, even when other evidence is not now observable. The upper basalt layers of older LIPs may have been removed by erosion or deformed by tectonic plate collisions occurring after the layer is formed. The is especially likely for earlier periods such as the Paleozoic and Proterozoic.

Giant dyke swarms having lengths >300 km are a common record of severely eroded LIPs. Both radial and linear dyke swarm configurations exist. Radial swarms with an areal extent >2000 km and linear swarms extending >1000 km are known. The linear dyke swarms often have a high proportion of dykes relative to country rocks, particularly when the width of the linear field is less than 100 km. The dykes have a typical width of 20–100 meters, although ultramafic dykes with widths greater than 1000 meters have been reported.

Dykes are typically sub-vertical to vertical. When upward flowing (dyke-forming) magma encounters horizontal boundaries or weaknesses, such as between layers in a sedimentary deposit, the magma can flow horizontally creating a sill. Some sill provinces have areal extents >1000 km.

Correlations with LIP formation

Correlation with hot-spots

The early volcanic activity of major hotspots, postulated to result from deep mantle plumes, is frequently accompanied by flood basalts. These flood basalt eruptions have resulted in large accumulations of basaltic lavas emplaced at a rate greatly exceeding that seen in contemporary volcanic processes. Continental rifting commonly follows flood basalt volcanism. Flood basalt provinces may also occur as a consequence of the initial hot-spot activity in ocean basins as well as on continents. It is possible to track the hot spot back to the flood basalts of a large igneous province; the table below correlates large igneous provinces with the track of a specific hot spot.

Province	Region	Hotspot	Reference
Columbia River Basalt	Northwestern USA	Yellowstone hotspot	
Ethiopia-Yemen Flood Basalts	Ethiopia, Yemen		
North Atlantic Igneous Province	Northern Canada, Greenland, the Faeroe Islands, Norway, Ireland and Scotland	Iceland hotspot	
Deccan Traps	India & southern Pakistan	Réunion hotspot	
Rahjamal Traps	Eastern India	Ninety East Ridge	
Kerguelen Plateau	Indian Ocean	Kerguelen hotspot	
Ontong-Java Plateau	Pacific Ocean	Louisville hotspot	
Paraná and Etendeka traps	Brazil–Namibia	Tristan hotspot	
Karoo-Ferrar Province	South Africa, Antarctica, Australia & New Zealand	Marion Island	
Caribbean large igneous province	Caribbean-Colombian oceanic plateau	Galápagos hotspot	
Mackenzie Large Igneous Province	Canadian Shield	Mackenzie hotspot	

Relationship to Extinction Events

Eruptions or emplacements of LIPs appear to have, in some cases, occurred simultaneously with oceanic anoxic events and extinction events. The most important examples are the Deccan Traps (Cretaceous–Paleogene extinction event), the Karoo-Ferrar (Pliensbachian-Toarcian extinction), the Central Atlantic Magmatic Province (Triassic-Jurassic extinction event), and the Siberian traps (Permian-Triassic extinction event).

Several mechanisms are proposed to explain the association of LIPs with extinction events. The eruption of basaltic LIPs onto the earth›s surface releases large volumes of sulfate gas, which forms sulfuric acid in the atmosphere; this absorbs heat and causes substantial cooling (e.g., the Laki eruption in Iceland, 1783). Oceanic LIPs can reduce oxygen in seawater by either direct oxidation reactions with metals in hydrothermal fluids or by causing algal blooms that consume large amounts of oxygen.

Ore Deposits

Large igneous provinces are associated with a handful of ore deposit types including:

- Ni–Cu PGEs
- Porphyries
- Iron oxide copper gold (IOCG)
- Kimberlites

Examples

There are a number of well documented examples of large igneous provinces identified by geological research.

Province	Region	Age (Ma)	Area ($10^6 km^2$	Volume ($10^6 km^3$	Also known as or includes	Reference
Agulhas Plateau	Southwest Indian Ocean, South Atlantic Ocean, Southern Ocean	140-95	0.3	1.2	Southeast African LIP Mozambique Ridge, Northeast Georgia Rise, Maud Rise, Astrid Ridge	
Columbia River Basalt	Northwestern US	17–6	0.16	0.175		
Ethiopia-Yemen Flood Basalts	Yemen–Ethiopia	31–25	0.6	0.35	Ethiopia	
North Atlantic Igneous Province	Northern Canada, Greenland, the Faeroe Islands, Norway, Ireland, and Scotland	62–55	1.3	6.6	High Arctic Large Igneous Province Jameson Land Thulean Plateau	

Province	Region	Age (Ma)	Area ($10^6 km^2$	Volume ($10^6 km^3$	Also known as or includes	Reference
Deccan Traps	India southern Pakistan	66	0.5–0.8	0.5–1.0		
Madagascar		88				
Rajmahal Traps		116				
Ontong-Java Plateau	Pacific Ocean	~122	1.86	8.4	Manihiki Plateau and Hikurangi Plateau	
Paraná and Etendeka traps	Brazil–Namibia	134–129	1.5	>1	Equatorial Atlantic Magmatic Province Brazilian Highlands	
Karoo-Ferrar Province	South Africa, Antarctica, Australia, and New Zealand	183–180	0.15–2	0.3		
Central Atlantic magmatic province	Northern South America, Northwest Africa, Iberia, Eastern North America	199–197	11	2.5 (2.0 – 3.0)		
Siberian Traps	Russia	250	1.5–3.9	0.9–2.0		
Emeishan Traps	Southwestern China	253–250	0.25	~0.3		
Warakurna large igneous province	Australia	1078–1073	1.5		Eastern Pilbara	

Large rhyolitic provinces (LRPs)

These LIPs are composed dominantly of felsic materials. Examples include:

- Whitsunday
- Sierra Madre Occidental (Mexico)
- Malani
- Chon Aike (Argentina)
- Gawler (Australia)

Large Andesitic Provinces (LAPs)

These LIPs are comprised dominantly of andesitic materials. Examples include:

- Island arcs such as Indonesia and Japan
- Active continental margins such as the Andes and the Cascades

- Continental collision zones such as the Anatolia-Iran zone

Large basaltic provinces (LBPs)

This subcategory includes most of the provinces included in the original LIP classifications. It is composed of continental flood basalts, oceanic flood basalts, and diffuse provinces.

Continental flood basalts

- Ethiopia-Yemen Continental Flood Basalts
- Columbia River Basalt Group
- Deccan Traps (India)
- Coppermine River Group (Canadian Shield)
- Equatorial Atlantic Magmatic Province (Maranhão, Brazil-Ghana)
- Midcontinent Rift System, Great Lakes Region, North America
- Paraná and Etendeka traps (Paraná, Brazil–NE Namibia)
- Brazilian Highlands
- Río de la Plata Craton (Uruguay)
- Karoo-Ferrar (South Africa–Antarctica)
- Siberian Traps (Russia)
- Emeishan Traps (western China)
- Central Atlantic Magmatic Province (eastern United States and Canada, northern South America, northwest Africa)
- North Atlantic Igneous Province (includes basalts in Greenland, Iceland, Ireland, Scotland, and Faroes)
- High Arctic Large Igneous Province (includes the Ellesmere Island Volcanics, Strand Fiord Formation, Alpha Ridge, Franz Josef Land, and Svalbard)

Oceanic flood basalts/ oceanic plateaus

- Wrangellia Terrane (Alaska and Canada)
- Caribbean large igneous province (Caribbean Sea)
- Kerguelen Plateau (Indian Ocean)
- Ontong Java Plateau, Manihiki Plateau and Hikurangi Plateau (southwest Pacific Ocean)
- Jameson Land

Large Basaltic–Rhyolitic Provinces (LBRPs)

- Snake River Plain – Oregon High Lava Plains
- Dongargarh (Ethiopia)

Large Plutonic Provinces (LPP)

Large granitic provinces (LGP)

- Patagonia
- Peru–Chile Batholith
- Coast Range Batholith (NW US)

Other Large Plutonic Provinces

- Parts of Central Atlantic Magmatic Province (eastern United States and Canada, northern South America, northwest Africa)

Related Structures

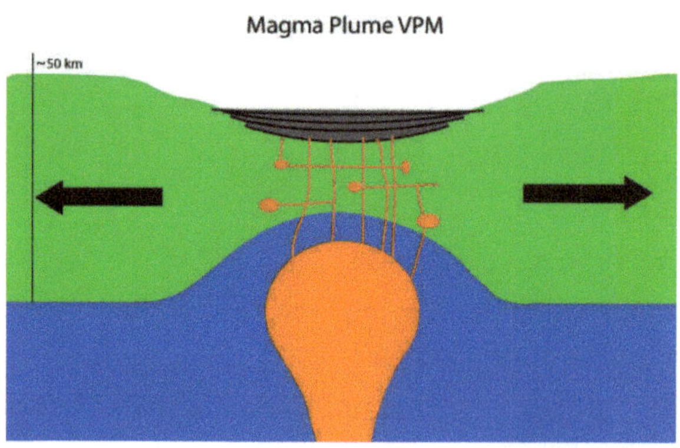

Extension thins the crust. Magma reaches the surface through radiating sills and dikes, forming basalt flows, as well as deep and shallow magma chambers below the surface. The crust gradually thins due to thermal subsidence, and originally horizontal basalt flows are rotated to become seaward dipping reflectors.

Volcanic rifted margins are found on the boundary of large igneous provinces. Volcanic margins form when rifting is accompanied by significant mantle melting, with volcanism occurring before and/or during continental breakup. Volcanic rifted margins are characterized by: a transitional crust composed of basaltic igneous rocks, including lava flows, sills, dikes, and gabbros, high volume basalt flows, seaward-dipping reflector sequences (SDRS) of basalt flows that were rotated during the early stages of breakup, limited passive-margin subsidence during and after breakup, and the presence of a lower crust with anomalously high seismic P-wave velocities in lower crustal bodies (LCBs), indicative of lower temperature, dense media.

Example of volcanic margins include:

- The Yemen margin

- The East Australian margin
- The West Indian margin
- The Hatton–Rockal margin
- The US East Coast
- The mid-Norwegian margin
- The Brazilian margins
- The Namibian margin

Dike Swarms

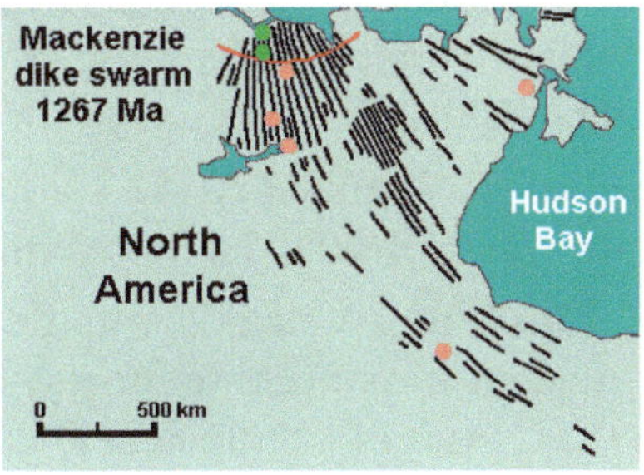

Map of the Mackenzie dike swarm in Canada

A dike swarm is a large geological structure consisting of a major group of parallel, linear, or radially oriented dikes intruded within continental crust. They consist of several to hundreds of dikes emplaced more or less contemporaneously during a single intrusive event, and are magmatic and stratigraphic. Such dike swarms are the roots of a volcanic province. Examples include:

- Mackenzie dike swarm (Canadian Shield)
- Long Range dikes (Newfoundland and Labrador, Canada)
- Mistassini dike swarm (western Quebec, Canada)
- Matachewan dike swarm (northern Ontario, Canada)
- Sorachi Plateau and Belt (Hokkaido, Japan)
- Uralian dike swarm (Russia)

Sills

A series of related sills that were formed essentially contemporaneously (within several million years) from related dikes comprise a LIP if their area is sufficiently large. Examples include:

- Winagami sill complex (northwestern Alberta, Canada)
- Bushveld Igneous Complex (South Africa) with an area of over 66,000 km^2 (25,000 sq mi), and a thickness reaching 9 kilometres (5.6 mi) thick.

REFERENCES

1. Foulger, G.R. (2010). Plates vs. Plumes: A Geological Controversy. Wiley-Blackwell. ISBN 978-1-4051-6148-0.
2. Sheth, Hetu C. (2007). ›Large Igneous Provinces (LIPs)›: Definition, recommended terminology, and a hierarchical classification (PDF). Earth-Science Reviews. 85 *(3–4): 117–124. Bibcode:2007ESRv...85..117S. doi:10.1016/j.earscirev.2007.07.005.*
3. Michael R. Rampino & Richard B. Stothers (1988). Flood Basalt Volcanism During the Past 250 Million Years (PDF). Science. 241 *(4866): 663–668. doi:10.1126/science.241.4866.663. PMID 17839077.*
4. Neal, C.; Mahoney, J.; Kroenke, L. (1997). The Ontong Java Plateau (PDF). Large Igneous Provinces: Continental, Oceanic, and Planetary Flood Volcanism, Geophysical Monograph 100.
5. David P.G. Bond & Paul B. Wignall (2014). Large igneous provinces and mass extinctions: An update. GSA Special Papers: 590. doi:10.1130/2014.2505(02).
6. Coffin, M; Eldholm, O (1992). Volcanism and continental break-up: a global compilation of large igneous provinces. In Storey, B.C., Alabaster, T., Pankhurst, R.J. Magmatism and the Causes of Continental Breakup. Special Publications. 68. London: Geological Society of London. pp. 17–30. Bibcode:1992GSLSP..68...17C. doi:10.1144/GSL.SP.1992.068.01.02.
7. Bryan, Scott; Ernst, Richard (2007). Proposed Revision to Large Igneous Province Classification. doi:10.1016/j.earscirev.2007.08.008.
8. N. I. Eremin; Platform Magmatism: Geology and Minerageny; ISSN 1075-7015, Geology of Ore Deposits, 2010, Vol. 52, No. 1, pp. 77–80. © Pleiades Publishing, Ltd., 2010. doi:10.1134/S1075701510010071
9. M.-F. Zhou et al.; Two magma series and associated ore deposit types in the Permian Emeishan large igneous province, SW China; Lithos; Vol. 103 (2008) pp.352–368; doi:10.1016/j.lithos.2007.10.006
10. Braun, J; The many surface expressions of mantle dynamics; Nature Geosciences; VOL 3; DECEMBER 2010; doi:10.1038/ngeo1020
11. Allen, Philip A (2011). Geodynamics: Surface impact of mantle processes. Nature Geoscience. 4: 498–499. doi:10.1038/ngeo1216.
12. Humphreys, Eugene; Schmandt, Brandon (2011). Looking for mantle plumes. Physics Today. 64 *(8): 34. doi:10.1063/PT.3.1217.*
13. Vincent Courtillot, Anne Davaille, Jean Besse, & Joann Stock ;Three distinct types of hotspots in the Earth's mantle; Earth and Planetary Science Letters; V. 205; 2003; pp.295–308
14. E. Humphreys and B. Schmandt; Looking for Mantle Plumes; Physics Today; August 2011; pp. 34- 39
15. J.T. Hagstrum; Antipodal hotspots and bipolar catastrophes: Were oceanic large-body impacts the cause?; Earth and Planetary Science Letters; Vol 236; (2005)

pp. 13–27; doi:10.1016/j.epsl.2005.02.020

16. Matthew G. Jackson & Richard W. Carlson; An ancient recipe for flood-basalt genesis; Nature (2011) doi:10.1038/nature10326
17. Coffin, M.F., Eldholm, O. (Eds.), 1991. Large Igneous Provinces: JOI/USSAC workshop report. The University of Texas at Austin Institute for Geophysics Technical Report, p. 114.
18. Coffin, M.F., Eldholm, O., 1992. Volcanism and continental break-up: a global compilation of large igneous provinces. In: Storey, B.C., Alabaster, T., Pankhurst, R.J. (Eds.), Magmatism and the Causes of Continental Break-up. Geological Society of London Special Publication, vol. 68, pp. 17–30.
19. Coffin, M.F., Eldholm, O., 1994. Large igneous provinces: crustal structure, dimensions, and external consequences. Reviews of Geophysics Vol. 32, pp. 1–36.
20. S.E. Bryan & R.E. Ernst; Revised definition of Large Igneous Provinces (LIPs); Earth-Science Reviews Vol. 86 (2008) pp. 175–202
21. Ernst, R. E.; Buchan, K. L., (1997), Giant Radiating Dyke Swarms: Their Use in Identifying Pre-Mesozoic Large Igneous Provinces and Mantle Plumes, in Mahoney, J. J.; Coffin, M. F. (editors), Large Igneous Provinces: Continental, Oceanic and Flood Volcanism (Geophysical Monograph 100), Washington D.C.: American Geophysical Union, p. 297, ISBN 0-87590-082-8
22. M.A. Richards, R.A. Duncan, V.E. Courtillot; Flood Basalts and Hot-Spot Tracks: Plume Heads and Tails; SCIENCE, VOL. 246 (1989) 103–108
23. Antretter, M.; Riisager, P.; Hall, S.; Zhao, X.; and Steinberger, B. (2004). Modelled palaeolatitudes for the Louisville hot spot and the Ontong Java Plateau, in Origin and Evolution of the Ontong Java Plateau Geological Society, London, Special Publications, v. 229, p. 21-30. doi:10.1144/GSL.SP.2004.229.01.03.
24. Barbara P. Nash, Michael E. Perkins, John N. Christensen, Der-Chuen Lee, & A.N. Halliday. "The Yellowstone hotspot in space and time: Nd and Hf isotopes in silicic magmas." Earth and Planetary Science Letters 247 (2006) 143–156
25. Weis, D.; et al. (1993). The Influence of Mantle Plumes in Generation of Indian Oceanic Crust. Geophysical Monograph. 70: 57–89. Bibcode:1992GMS....70...57W. doi:10.1029/gm070p0057.
26. E.V. Verzhbitsky. Geothermal regime and genesis of the Ninety-East and Chagos-Laccadive ridges. Journal of Geodynamics, Volume 35, Issue 3, April 2003, Pages 289–302
27. Sur l'âge des trapps basaltiques (On the ages of flood basalt events); Vincent E. Courtillot & Paul R. Renne; Comptes Rendus Geoscience; Vol: 335 Issue: 1, January 2003; pp: 113–140
28. Kaj Hoernle, Folkmar Hauff and Paul van den Bogaard, 70 m.y. history (139–69 Ma) for the Caribbean large igneous province, Geology; August 2004; v. 32; no. 8; p. 697-700 Abstract
29. Ernst, Richard E.; Buchan, Kenneth L. (2001). Mantle plumes: their identification through time. Geological Society of America. pp. 143, 145, 146, 147, 148, 259.

ISBN 0-8137-2352-3.

30. Kerr, AC (December 2005). Oceanic LIPS: Kiss of death. Elements. 1 *(5): 289–292. doi:10.2113/gselements.1.5.289.*

31. Gohl, K.; Uenzelmann-Neben, G.; Grobys, N. (2011). Growth and dispersal of a southeast African Large Igenous Province (PDF). South African Journal of Geology. 114 *(3-4): 379–386. doi:10.2113/gssajg.114.3-4.379. Retrieved April 2015.* Check date values in: |access-date= (help)

32. P.S. Ross, I. Ukstins Peateb, M.K. McClintocka, Y.G. Xuc, I.P. Skillingd, J.D.L. Whitea, and B.F. Houghtone. Mafic volcaniclastic deposits in flood basalt provinces: A review. Journal of Volcanology and Geothermal Research 145 (2005) 281–314

33. TH Torsvik, RD Tucker, LD Ashwal, EA Eide, NA Rakotosolofo, MJ de Wit. "Late Cretaceous magmatism in Madagascar: palaeomagnetic evidence for a stationary Marion hotspot." Earth and Planetary Science Letters, Volume 164, Issues 1–2, 15 December 1998, Pages 221–232

34. Knight, K.B.; Nomade S.; Renne P.R.; Marzoli A.; Bertrand H.; Youbi N. (2004). The Central Atlantic Magmatic Province at the Triassic–Jurassic boundary: paleomagnetic and 40Ar/39Ar evidence from Morocco for brief, episodic volcanism. Earth and Planetary Science Letters. Elsevier. 228 *(1–2): 143–160. Bibcode:2004E&PSL.228..143K. doi:10.1016/j.epsl.2004.09.022. Retrieved 21 September 2012.*

35. Blackburn, Terrence J.; Olsen, Paul E.; Bowring, Samuel A.; McLean, Noah M.; Kent, Dennis V; Puffer, John; McHone, Greg; Rasbury, Troy; Et-Touhami, Mohammed (2013). Zircon U–Pb Geochronology Links the End-Triassic Extinction with the Central Atlantic Magmatic Province. Science. 340 *(6135): 941–945. Bibcode:2013Sci...340..941B. doi:10.1126/science.1234204. PMID 23519213.*

36. MTD Wingate, F Pirajno, & PA Morris. Warakurna large igneous province: A new Mesoproterozoic large igneous province in west-central Australia. Geology (2004) Volume 32; pp.105–108 doi:10.1130/G20171.1

37. Puchkov, Victor; Ernst, Richard E.; Hamilton, Michael A.; Söderlund, Ulf; Sergeeva, Nina (2016). A Devonian >2000-km-long dolerite dyke swarm-belt and associated basalts along the Urals-Novozemelian fold-belt: part of an East-European (Baltica) LIP tracing the Tuzo Superswell. GFF. 138: 6–16. doi:10.1080/11035897.2015.1118406.

38. Tony Waltham (2009). Foundations of Engineering Geology (3rd ed.). Taylor & Francis. p. 20. ISBN 0-415-46959-7. Philip Kearey; Keith A. Klepeis; Frederick J. Vine (2009). Chapter 10: Orogenic belts. Global Tectonics (3rd ed.). Wiley-Blackwell. p. 287. ISBN 1-4051-0777-4.

39. Chambers 21st Century Dictionary. Allied Publishers. 1999. p. 972. ISBN 978-0550106254. Retrieved 27 June 2012.

40. Friedman G.M. (1994). Pangean Orogenic and Epeirogenic Uplifts and Their Possible Climatic Significance. In Klein G.O. Pangea: Paleoclimate, Tectonics, and Sedimentation During Accretion, Zenith, and Breakup of a Supercontinent.

Geological Society of America Special Paper. 288. p. 160. ISBN 9780813722887.

41. N. H. Woodcock; Robin A. Strachan (2000). Chapter 12: The Caledonian Orogeny: a multiple plate collision. Geological History of Britain and Ireland. Wiley-Blackwell. p. 202, Figure 12.11. ISBN 0-632-03656-7.

42. Frank Press (2003). Understanding Earth (4th ed.). Macmillan. pp. 468–469. ISBN 0-7167-9617-1.

43. Gerald Schubert; Donald Lawson Turcotte; Peter Olson (2001). §2.5.4 Why are island arcs arcs?. Mantle Convection in the Earth and Planets. Cambridge University Press. pp. 35–36. ISBN 0-521-79836-1.

44. PA Allen (1997). Isostasy in zones of convergence. Earth Surface Processes. Wiley-Blackwell. p. 36 ff. ISBN 0-632-03507-2.

45. Gerard V. Middleton; Peter R. Wilcock (1994). §5.5 Isostasy. Mechanics in the Earth and Environmental Sciences (2nd ed.). Cambridge University Press. p. 170. ISBN 0-521-44669-4.

46. DeCelles P.G. & Giles K.A. (1996). Foreland basin systems (PDF). Basin Research. 8 *(2): 105–123. doi:10.1046/j.1365-2117.1996.01491.x.*

47. David Johnson (2004). The orogenic cycle. The geology of Australia. Cambridge University Press. p. 48 ff. ISBN 0-521-84121-6.

48. In other words, orogeny is only a phase in the existence of an orogen. Five characteristics of the orogenic cycle are listed by Robert J. Twiss; Eldridge M. Moores (1992). Plate tectonic models of orogenic core zones. Structural Geology (2nd ed.). Macmillan. p. 493. ISBN 0-7167-2252-6.

49. N. H. Woodcock; Robin A. Strachan (2000). Chapter 12: The Caledonian Orogeny: A Multiple Plate Collision. cited work. p. 187 ff. ISBN 0-632-03656-7.

50. Olivier Merle (1998). §1.1 Nappes, overthrusts and fold-nappes. Emplacement Mechanisms of Nappes and Thrust Sheets. Petrology and Structural Geology. 9. Springer. p. 1 ff. ISBN 0-7923-4879-6.

51. For example, see Patrick L Osborne (2000). Tropical Ecosystems and Ecological Concepts. Cambridge University Press. p. 11. ISBN 0-521-64523-9. Continental drift and plate tectonics help to explain both the similarities and the differences in the distribution of plants and animals over the continents and John C Briggs (1 987). Biogeography and Plate Tectonics. Elsevier. p. 131. ISBN 0-444-42743-0. It will not be possible to construct a thorough account of the history of the southern hemisphere without the evidence from both the biological and the earth sciences Check date values in: |date= (help)

52. Paul D. Lowman (2002). Chapter 7: Geology and biology: the influence of life on terrestrial geology. Exploring Space, Exploring Earth: New Understanding of the Earth from Space Research. Cambridge University Press. pp. 286–287. ISBN 0-521-89062-4.

53. Seema Sharma (2005). Atmosphere: origin. Encyclopaedia of Climatology. Anmol Publications PVT. LTD. p. 30 ff. ISBN 81-261-2442-3.

54. Richard J. Huggett (2007). Fundamentals of Geomorphology (2nd ed.). Routledge. p. 104. ISBN 0-415-39084-2.

55. Gerhard Einsele (2000). Sedimentary Basins: Evolution, Facies, and Sediment Budget (2nd ed.). Springer. p. 453. ISBN 3-540-66193-X. Without denudation, even relatively low uplift rates as characteristic of epeirogenetic movements (e.g. 20m/MA) would generate highly elevated regions in geological time periods.

56. Ian Douglas; Richard John Huggett; Mike Robinson (2002). Companion Encyclopedia of Geography: The Environment and Humankind. Taylor & Francis. p. 33. ISBN 0-415-27750-7.

57. Peter J Coney (1970). The Geotectonic Cycle and the New Global Tectonics. Geological Society of America Bulletin. 81 *(3): 739–748. Bibcode:1970GSAB...81..739C. doi:10.1130/0016-7606(1970)81[739:TGCATN]2.0.CO;2.*

58. Lee, C.-T.; Yin, Q; Rudnick, RL; Chesley, JT; Jacobsen, SB (2000). Osmium Isotopic Evidence for Mesozoic Removal of Lithospheric Mantle Beneath the Sierra Nevada, California (PDF). Science. 289 *(5486): 1912–6. Bibcode:2000Sci...289.1912L. doi:10.1126/science.289.5486.1912. PMID 10988067.*

59. John Gerrard (1990). Mountain Environments: An Examination of the Physical Geography of Mountains. MIT Press. p. 9. ISBN 0-262-07128-2.

60. Manley, Curtis R.; Glazner, Allen F.; Farmer, G. Lang (2000). Timing of Volcanism in the Sierra Nevada of California: Evidence for Pliocene Delamination of the Batholithic Root?. Geology. 28 *(9): 811. Bibcode:2000Geo....28..811M. doi:10.1130/0091-7613(2000)28<811:TOVITS>2.0.CO;2.*

61. Arthur Holmes; Doris L. Holmes (2004). Holmes Principles of Physical Geology (4th ed.). Taylor & Francis. p. 92. ISBN 0-7487-4381-2.

62. The Formation of the Rocky Mountains, Mountains in Nature, n.d., retrieved 29 January 2014

63. Zwart, H. J., 1967. The duality of orogenicb elts. Geol. Mijnbouw. 46, 283-309 (referred in Pitcher 1979)

64. W.S. Pitcher, "The nature, ascent and emplacement of granitic magmas", Journal of the Geological Society 1979; v. 136; p. 627-662

65. de Beaumont, J.-B., Élie (1852 (English synopsis in Dennis (1982))). Notice sur les Systèmes de Montagnes (Note on Mountain Systems). Bertrand, Paris. pp. 1543 pp. Check date values in: |date= (help)

CHAPTER 5

Plate Tectonics

Chapter Contents

- **Introduction**
- **Key Principles**
- **Types of Plate Boundaries**
- **Driving Forces of Plate Motion**
- **Development of the Theory**
- **Plate Reconstruction**
- **Current Plates**

Plate tectonics, theory dealing with the dynamics of Earth's outer shell, the lithosphere, that revolutionized Earth sciences by providing a uniform context for understanding mountain-building processes, volcanoes, and earthquakes, as well as understanding the evolution of Earth's surface and reconstructing its past continental and oceanic configurations.

INTRODUCTION

Plate tectonics (from the Late Latin *tectonicus*, from the Greek: τεκτονικός "pertaining to building") is a scientific theory describing the large-scale motion of 7 large plates and the movements of a larger number of smaller plates of the Earth›s lithosphere, over the last hundreds of millions of years. The theoretical model builds on the concept of continental drift developed during the first few decades of the 20th century. The geoscientific community accepted plate-tectonic theory after seafloor spreading was validated in the late 1950s and early 1960s.

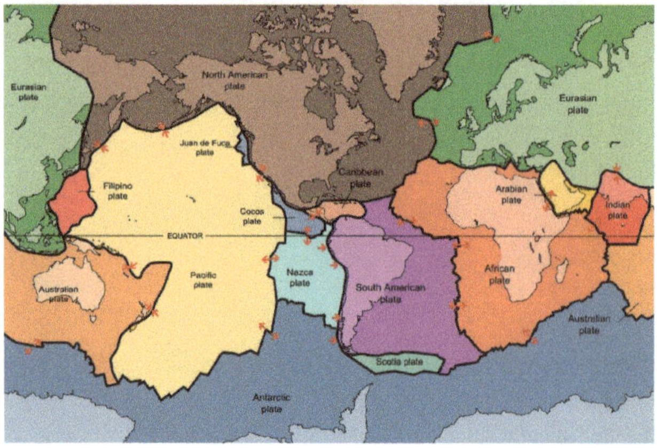

The tectonic plates of the world were mapped in the second half of the 20th century.

The lithosphere, which is the rigid outermost shell of a planet (the crust and upper mantle), is broken up into tectonic plates. The Earth›s lithosphere is composed of seven or eight major plates (depending on how they are defined) and many minor plates. Where the plates meet, their relative motion determines the type of boundary: convergent, divergent, or transform. Earthquakes, volcanic activity, mountain-building, and oceanic trench formation occur along these plate boundaries. The relative movement of the plates typically ranges from zero to 100 mm annually.

Tectonic plates are composed of oceanic lithosphere and thicker continental lithosphere, each topped by its own kind of crust. Along convergent boundaries, subduction carries plates into the mantle; the material lost is roughly balanced by the formation of new (oceanic) crust along divergent margins by seafloor spreading. In this way, the total surface of the lithosphere remains the same. This prediction of plate tectonics is also referred to as the conveyor belt principle. Earlier theories (that still have some supporters) propose gradual shrinking (contraction) or gradual expansion of the globe.

Tectonic plates are able to move because the Earth›s lithosphere has greater strength than the underlying asthenosphere. Lateral density variations in the mantle result in convection. Plate movement is thought to be driven by a combination of the motion of the seafloor away from the spreading ridge (due to variations in topography and density of the crust, which result in differences in gravitational forces) and drag, with downward suction, at the subduction zones. Another explanation lies in the different forces generated by tidal forces of the Sun and Moon. The relative importance of each of these factors and their relationship to each other is unclear, and still the subject of much debate.

KEY PRINCIPLES

The outer layers of the Earth are divided into the lithosphere and asthenosphere. This is based on differences in mechanical properties and in the method for the transfer of heat. Mechanically, the lithosphere is cooler and more rigid, while the asthenosphere is hotter and flows more easily. In terms of heat transfer, the lithosphere loses heat by conduction, whereas the asthenosphere also transfers heat by convection and has a nearly adiabatic temperature gradient. This division should not be confused with the *chemical* subdivision of these same layers into the mantle (comprising both the asthenosphere and the mantle portion of the lithosphere) and the crust: a given piece of mantle may be part of the lithosphere or the asthenosphere at different times depending on its temperature and pressure.

The key principle of plate tectonics is that the lithosphere exists as separate and distinct *tectonic plates*, which ride on the fluid-like (visco-elastic solid) asthenosphere. Plate motions range up to a typical 10–40 mm/year (Mid-Atlantic Ridge; about as fast as fingernails grow), to about 160 mm/year (Nazca Plate; about as fast as hair grows). The driving mechanism behind this movement is described below.

Tectonic lithosphere plates consist of lithospheric mantle overlain by either or both of two types of crustal material: oceanic crust (in older texts called *sima* from silicon and magnesium) and continental crust (*sial* from silicon and aluminium). Average oceanic lithosphere is typically 100 km (62 mi) thick; its thickness is a function of its age: as time passes, it conductively cools and subjacent cooling mantle is added to its base. Because it is formed at mid-ocean ridges and spreads outwards, its thickness is therefore a function of its distance from the mid-ocean ridge where it was formed. For a typical distance that oceanic lithosphere must travel before being subducted, the thickness varies from about 6 km (4 mi) thick at mid-ocean ridges to greater than 100 km (62 mi) at subduction zones; for shorter or longer distances, the subduction zone (and therefore also the mean) thickness becomes smaller or larger, respectively. Continental lithosphere is typically ~200 km thick, though this varies considerably between basins, mountain ranges, and stable cratonic interiors of continents. The two types of crust also differ in thickness, with continental crust being considerably thicker than oceanic (35 km vs. 6 km).

The location where two plates meet is called a *plate boundary*. Plate boundaries are commonly associated with geological events such as earthquakes and the creation of topographic features such as mountains, volcanoes, mid-ocean ridges, and oceanic trenches. The majority of the world›s active volcanoes occur along plate boundaries, with the Pacific Plate›s Ring of Fire being the most active and widely known today. These boundaries are discussed in further detail below. Some volcanoes occur in the interiors of plates, and these have been variously attributed to internal plate deformation and to mantle plumes. As explained above, tectonic plates may include continental crust or oceanic crust, and most plates contain both. For example, the African Plate includes the continent and parts of the floor of the Atlantic and Indian Oceans. The distinction between oceanic crust and continental crust is based on their modes of formation. Oceanic crust is formed at sea-floor spreading centers, and continental crust is formed through arc volcanism and accretion of terranes through tectonic processes, though some of these terranes may contain ophiolite sequences, which are pieces of oceanic crust considered to be part of the continent when they exit the standard cycle of formation and spreading centers and subduction beneath continents. Oceanic crust is also denser than continental crust owing to their different compositions. Oceanic crust is denser because it has less silicon and more heavier elements (mafic) than continental crust (felsic). As a result of this density stratification, oceanic crust generally lies below sea level (for example most of the Pacific Plate), while continental crust buoyantly projects above sea level (see the page isostasy for explanation of this principle).

TYPES OF PLATE BOUNDARIES

Three types of plate boundaries exist, with a fourth, mixed type, characterized by the way the plates move relative to each other. They are associated with different types of surface phenomena. The different types of plate boundaries are:

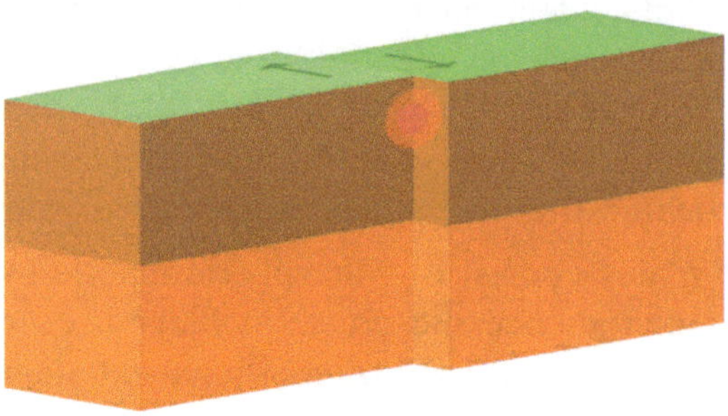

Transform boundary

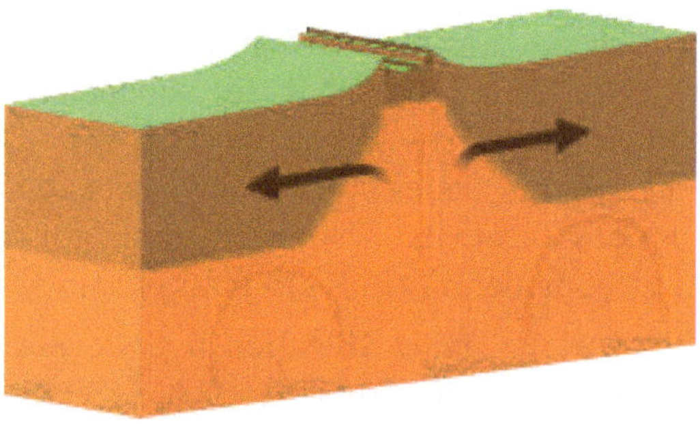

Divergent boundary

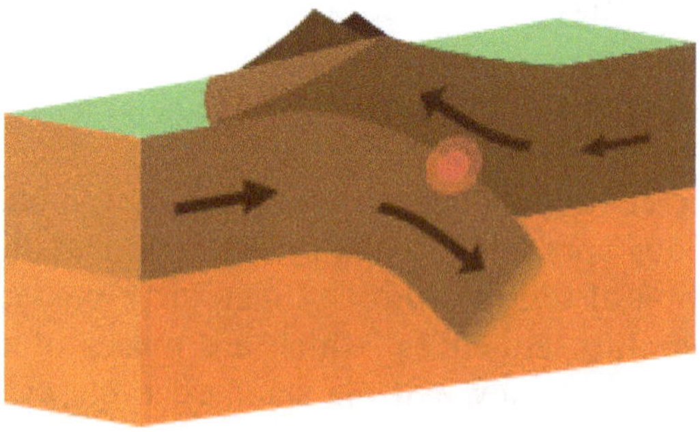

Convergent boundary

- *Transform boundaries (Conservative)* occur where two lithospheric plates slide, or perhaps more accurately, grind past each other along transform faults, where plates are neither created nor destroyed. The relative motion of the two plates is either sinistral (left side toward the observer) or dextral (right side toward the observer). Transform faults occur across a spreading center. Strong earthquakes can occur along a fault. The San Andreas Fault in California is an example of a transform boundary exhibiting dextral motion.

- *Divergent boundaries (Constructive)* occur where two plates slide apart from each other. At zones of ocean-to-ocean rifting, divergent boundaries form by seafloor spreading, allowing for the formation of new ocean basin. As the ocean plate splits, the ridge forms at the spreading center, the ocean basin expands, and finally, the plate area increases causing many small volcanoes and/or shallow earthquakes. At zones of continent-to-continent rifting, divergent boundaries may cause new ocean basin to form as the continent splits, spreads, the central rift collapses, and ocean fills the basin. Active zones of Mid-ocean ridges (e.g., Mid-Atlantic Ridge and East Pacific Rise), and continent-to-continent rifting (such as Africa›s East African Rift and Valley, Red Sea) are examples of divergent boundaries.
- *Convergent boundaries (Destructive)* (or *active margins*) occur where two plates slide toward each other to form either a subduction zone (one plate moving underneath the other) or a continental collision. At zones of ocean-to-continent subduction (e.g. the Andes mountain range in South America, and the Cascade Mountains in Western United States), the dense oceanic lithosphere plunges beneath the less dense continent. Earthquakes trace the path of the downward-moving plate as it descends into asthenosphere, a trench forms, and as the subducted plate is heated it releases volatiles, mostly water from hydrous minerals, into the surrounding mantle. The addition of water lowers the melting point of the mantle material above the subducting slab, causing it to melt. The magma that results typically leads to volcanism.At zones of ocean-to-ocean subduction (e.g. Aleutian islands, Mariana Islands, and the Japanese island arc), older, cooler, denser crust slips beneath less dense crust. This causes earthquakes and a deep trench to form in an arc shape. The upper mantle of the subducted plate then heats and magma rises to form curving chains of volcanic islands. Deep marine trenches are typically associated with subduction zones, and the basins that develop along the active boundary are often called foreland basins. Closure of ocean basins can occur at continent-to-continent boundaries (e.g., Himalayas and Alps): collision between masses of granitic continental lithosphere; neither mass is subducted; plate edges are compressed, folded, uplifted.
- *Plate boundary zones* occur where the effects of the interactions are unclear, and the boundaries, usually occurring along a broad belt, are not well defined and may show various types of movements in different episodes.

DRIVING FORCES OF PLATE MOTION

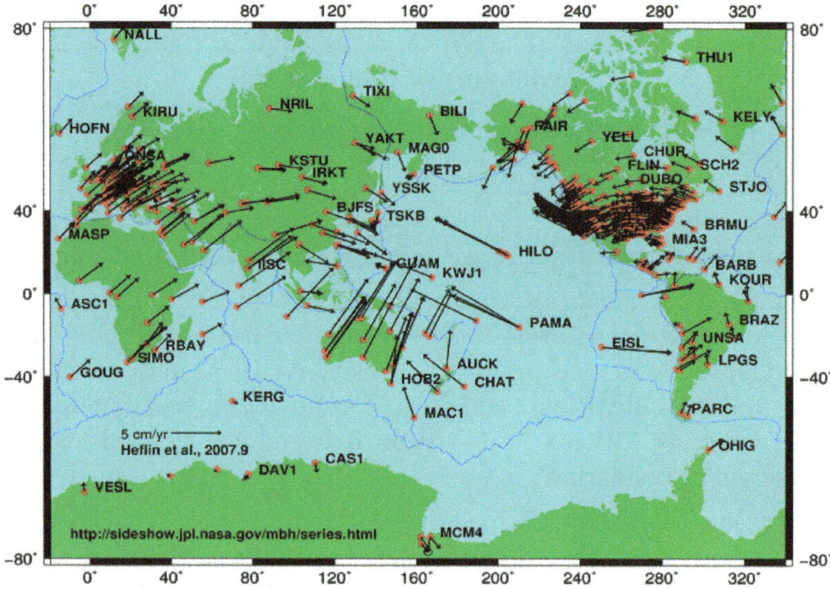

Plate motion based on Global Positioning System (GPS) satellite data from NASA JPL. The vectors show direction and magnitude of motion.

It has generally been accepted that tectonic plates are able to move because of the relative density of oceanic lithosphere and the relative weakness of the asthenosphere. Dissipation of heat from the mantle is acknowledged to be the original source of the energy required to drive plate tectonics through convection or large scale upwelling and doming. The current view, though still a matter of some debate, asserts that as a consequence, a powerful source of plate motion is generated due to the excess density of the oceanic lithosphere sinking in subduction zones. When the new crust forms at mid-ocean ridges, this oceanic lithosphere is initially less dense than the underlying asthenosphere, but it becomes denser with age as it conductively cools and thickens. The greater density of old lithosphere relative to the underlying asthenosphere allows it to sink into the deep mantle at subduction zones, providing most of the driving force for plate movement. The weakness of the asthenosphere allows the tectonic plates to move easily towards a subduction zone. Although subduction is thought to be the strongest force driving plate motions, it cannot be the only force since there are plates such as the North American Plate which are moving, yet are nowhere being subducted. The same is true for the enormous Eurasian Plate. The sources of plate motion are a matter of intensive research and discussion among scientists. One of the main points is that the kinematic pattern of the movement itself should be separated clearly from the possible geodynamic mechanism that is invoked as the driving force of the observed movement, as some patterns may be explained by more than one mechanism. In short, the driving forces advocated at the moment can be divided into three categories based on the relationship to the movement: mantle dynamics related, gravity related (mostly secondary forces).

Driving Forces Related to Mantle Dynamics

For much of the last quarter century, the leading theory of the driving force behind tectonic plate motions envisaged large scale convection currents in the upper mantle, which can be transmitted through the asthenosphere. This theory was launched by Arthur Holmes and some forerunners in the 1930s and was immediately recognized as the solution for the acceptance of the theory as originally discussed in the papers of Alfred Wegener in the early years of the century. However, despite its acceptance, it was long debated in the scientific community because the leading ("fixist") theory still envisaged a static Earth without moving continents up until the major breakthroughs of the early sixties.

Two- and three-dimensional imaging of Earth›s interior (seismic tomography) shows a varying lateral density distribution throughout the mantle. Such density variations can be material (from rock chemistry), mineral (from variations in mineral structures), or thermal (through thermal expansion and contraction from heat energy). The manifestation of this varying lateral density is mantle convection from buoyancy forces.

How mantle convection directly and indirectly relates to plate motion is a matter of ongoing study and discussion in geodynamics. Somehow, this energy must be transferred to the lithosphere for tectonic plates to move. There are essentially two main types of forces that are thought to influence plate motion: friction and gravity.

- Basal drag (friction): Plate motion driven by friction between the convection currents in the asthenosphere and the more rigid overlying lithosphere.
- Slab suction (gravity): Plate motion driven by local convection currents that exert a downward pull on plates in subduction zones at ocean trenches. Slab suction may occur in a geodynamic setting where basal tractions continue to act on the plate as it dives into the mantle (although perhaps to a greater extent acting on both the under and upper side of the slab).

Lately, the convection theory has been much debated as modern techniques based on 3D seismic tomography still fail to recognize these predicted large scale convection cells. Therefore, alternative views have been proposed:

In the theory of plume tectonics developed during the 1990s, a modified concept of mantle convection currents is used. It asserts that super plumes rise from the deeper mantle and are the drivers or substitutes of the major convection cells. These ideas, which find their roots in the early 1930s with the so-called "fixistic ideas of the European and Russian Earth Science Schools, find resonance in the modern theories which envisage hot spots/mantle plumes which remain fixed and are overridden by oceanic and continental lithosphere plates over time and leave their traces in the geological record (though these phenomena are not invoked as real driving mechanisms, but rather as modulators). Modern theories that continue building on the older mantle doming concepts and see plate movements as a secondary phenomena are beyond the scope of this page and are discussed elsewhere (for example on the plume tectonics page).

Another theory is that the mantle flows neither in cells nor large plumes but rather as a series of channels just below the Earth›s crust, which then provide basal friction to the lithosphere. This theory, called surge tectonics, became quite popular in geophysics and geodynamics during the 1980s and 1990s. Recent research, based on three-dimensional computer modeling, suggests that plate geometry is governed by a feedback between mantle convection patterns and the strength of the lithosphere.

Driving Forces Related to Gravity

Forces related to gravity are usually invoked as secondary phenomena within the framework of a more general driving mechanism such as the various forms of mantle dynamics described above.

Gravitational sliding away from a spreading ridge: According to many authors, plate motion is driven by the higher elevation of plates at ocean ridges. As oceanic lithosphere is formed at spreading ridges from hot mantle material, it gradually cools and thickens with age (and thus adds distance from the ridge). Cool oceanic lithosphere is significantly denser than the hot mantle material from which it is derived and so with increasing thickness it gradually subsides into the mantle to compensate the greater load. The result is a slight lateral incline with increased distance from the ridge axis.

This force is regarded as a secondary force and is often referred to as "ridge push". This is a misnomer as nothing is "pushing" horizontally and tensional features are dominant along ridges. It is more accurate to refer to this mechanism as gravitational sliding as variable topography across the totality of the plate can vary considerably and the topography of spreading ridges is only the most prominent feature. Other mechanisms generating this gravitational secondary force include flexural bulging of the lithosphere before it dives underneath an adjacent plate which produces a clear topographical feature that can offset, or at least affect, the influence of topographical ocean ridges, and mantle plumes and hot spots, which are postulated to impinge on the underside of tectonic plates.

Slab-pull: Current scientific opinion is that the asthenosphere is insufficiently competent or rigid to directly cause motion by friction along the base of the lithosphere. Slab pull is therefore most widely thought to be the greatest force acting on the plates. In this current understanding, plate motion is mostly driven by the weight of cold, dense plates sinking into the mantle at trenches. Recent models indicate that trench suction plays an important role as well. However, as the North American Plate is nowhere being subducted, yet it is in motion presents a problem. The same holds for the African, Eurasian, and Antarctic plates.

Gravitational sliding away from mantle doming: According to older theories, one of the driving mechanisms of the plates is the existence of large scale asthenosphere/ mantle domes which cause the gravitational sliding of lithosphere plates away from them. This gravitational sliding represents a secondary phenomenon of this basically vertically oriented mechanism. This can act on various scales, from the small scale of one island arc up to the larger scale of an entire ocean basin.

Driving Forces Related to Earth Rotation

Alfred Wegener, being a meteorologist, had proposed tidal forces and pole flight force as the main driving mechanisms behind continental drift; however, these forces were considered far too small to cause continental motion as the concept then was of continents plowing through oceanic crust. Therefore, Wegener later changed his position and asserted that convection currents are the main driving force of plate tectonics in the last edition of his book in 1929.

However, in the plate tectonics context (accepted since the seafloor spreading proposals of Heezen, Hess, Dietz, Morley, Vine, and Matthews (see below) during the early 1960s), the oceanic crust is suggested to be in motion *with* the continents which caused the proposals related to Earth rotation to be reconsidered. In more recent literature, these driving forces are:

- Tidal drag due to the gravitational force the Moon (and the Sun) exerts on the crust of the Earth
- Global deformation of the geoid due to small displacements of rotational pole with respect to the Earth's crust;
- Other smaller deformation effects of the crust due to wobbles and spin movements of the Earth rotation on a smaller time scale.

Forces that are small and generally negligible are:

- The Coriolis force
- The centrifugal force, which is treated as a slight modification of gravity

For these mechanisms to be overall valid, systematic relationships should exist all over the globe between the orientation and kinematics of deformation and the geographical latitudinal and longitudinal grid of the Earth itself. Ironically, these systematic relations studies in the second half of the nineteenth century and the first half of the twentieth century underline exactly the opposite: that the plates had not moved in time, that the deformation grid was fixed with respect to the Earth equator and axis, and that gravitational driving forces were generally acting vertically and caused only local horizontal movements (the so-called pre-plate tectonic, "fixist theories"). Later studies (discussed below on this page), therefore, invoked many of the relationships recognized during this pre-plate tectonics period to support their theories (see the anticipations and reviews in the work of van Dijk and collaborators).

Of the many forces discussed in this paragraph, tidal force is still highly debated and defended as a possible principle driving force of plate tectonics. The other forces are only used in global geodynamic models not using plate tectonics concepts (therefore beyond the discussions treated in this section) or proposed as minor modulations within the overall plate tectonics model.

In 1973, George W. Moore of the USGS and R. C. Bostrom presented evidence for a general westward drift of the Earth's lithosphere with respect to the mantle. He concluded that tidal forces (the tidal lag or "friction") caused by the Earth's rotation and the forces acting upon it by the Moon are a driving force for plate tectonics. As the Earth spins eastward beneath the moon, the moon's gravity ever so slightly pulls the Earth's surface layer back westward, just as proposed by Alfred Wegener (see above). In a more recent 2006 study, scientists reviewed and advocated these earlier proposed ideas. It has also been suggested recently in Lovett (2006) that this observation may also explain why Venus and Mars have no plate tectonics, as Venus has no moon and Mars' moons are too small to have significant tidal effects on the planet. In a recent paper, it was suggested that, on the other hand, it can easily be observed that many plates are moving north and eastward, and that the dominantly westward motion of the Pacific Ocean basins derives simply from the eastward bias of the Pacific spreading center (which is not a predicted manifestation of such lunar forces). In the same paper the authors admit, however, that relative to the lower mantle, there is a slight westward component in the motions of all the plates. They demonstrated though that the westward drift, seen only for the past 30 Ma, is attributed to the increased dominance of the steadily growing and accelerating Pacific plate. The debate is still open.

Relative Significance of each Driving Force Mechanism

The vector of a plate's motion is a function of all the forces acting on the plate; however, therein lies the problem regarding the degree to which each process contributes to the overall motion of each tectonic plate.

The diversity of geodynamic settings and the properties of each plate result from the impact of the various processes actively driving each individual plate. One method of dealing with this problem is to consider the relative rate at which each plate is moving as well as the evidence related to the significance of each process to the overall driving force on the plate.

One of the most significant correlations discovered to date is that lithospheric plates attached to downgoing (subducting) plates move much faster than plates not attached to subducting plates. The Pacific plate, for instance, is essentially surrounded by zones of subduction (the so-called Ring of Fire) and moves much faster than the plates of the Atlantic basin, which are attached (perhaps one could say ‹welded›) to adjacent continents instead of subducting plates. It is thus thought that forces associated with the downgoing plate (slab pull and slab suction) are the driving forces which determine the motion of plates, except for those plates which are not being subducted. The driving forces of plate motion continue to be active subjects of on-going research within geophysics and tectonophysics.

DEVELOPMENT OF THE THEORY

Summary

In line with other previous and contemporaneous proposals, in 1912 the meteorologist Alfred Wegener amply described what he called continental drift, expanded in his 1915 book *The Origin of Continents and Oceans* and the scientific debate started that would end up fifty years later in the theory of plate tectonics. Starting from the idea (also expressed by his forerunners) that the present continents once formed a single land mass (which was called Pangea later on) that drifted apart, thus releasing the continents from the Earth's mantle and likening them to "icebergs" of low density granite floating on a sea of denser basalt.Supporting evidence for the idea came from the dove-tailing outlines of South America›s east coast and Africa›s west coast, and from the matching of the rock formations along these edges. Confirmation of their previous contiguous nature also came from the fossil plants *Glossopteris* and *Gangamopteris*, and the therapsid or mammal-like reptile *Lystrosaurus*, all widely distributed over South America, Africa, Antarctica, India and Australia. The evidence for such an erstwhile joining of these continents was patent to field geologists working in the southern hemisphere. The South African Alex du Toit put together a mass of such information in his 1937 publication *Our Wandering Continents*, and went further than Wegener in recognising the strong links between the Gondwana fragments.

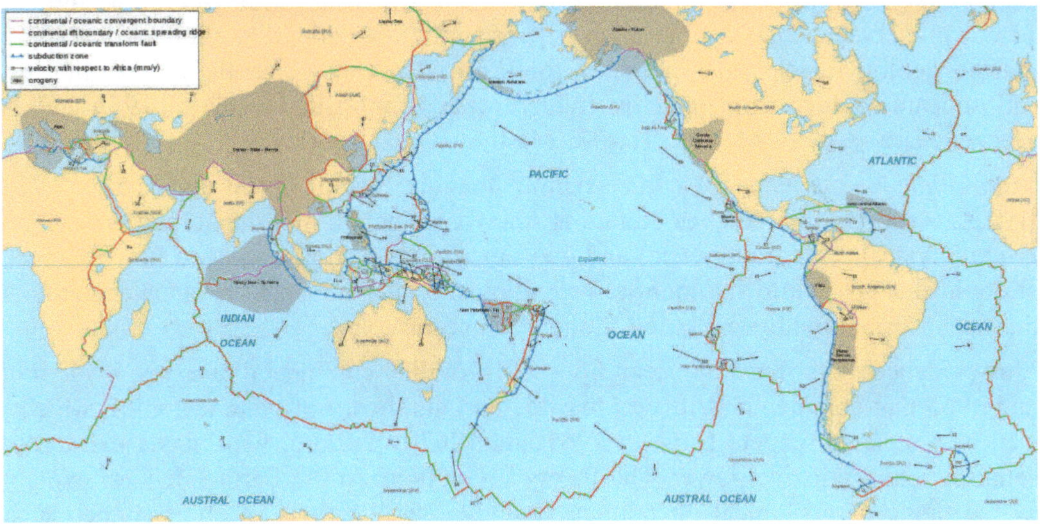

Detailed map showing the tectonic plates with their movement vectors.

But without detailed evidence and a force sufficient to drive the movement, the theory was not generally accepted: the Earth might have a solid crust and mantle and a liquid core, but there seemed to be no way that portions of the crust could move around. Distinguished scientists, such as Harold Jeffreys and Charles Schuchert, were outspoken critics of continental drift.

Despite much opposition, the view of continental drift gained support and a lively debate started between drifters or mobilists (proponents of the theory) and fixists (opponents). During the 1920s, 1930s and 1940s, the former reached important milestones proposing that convection currents might have driven the plate movements, and that spreading may have occurred below the sea within the oceanic crust. Concepts close to the elements now incorporated in plate tectonics were proposed by geophysicists and geologists (both fixists and mobilists) like Vening-Meinesz, Holmes, and Umbgrove.

One of the first pieces of geophysical evidence that was used to support the movement of lithospheric plates came from paleomagnetism. This is based on the fact that rocks of different ages show a variable magnetic field direction, evidenced by studies since the mid–nineteenth century. The magnetic north and south poles reverse through time, and, especially important in paleotectonic studies, the relative position of the magnetic north pole varies through time. Initially, during the first half of the twentieth century, the latter phenomenon was explained by introducing what was called polar wander (see apparent polar wander), i.e., it was assumed that the north pole location had been shifting through time. An alternative explanation, though, was that the continents had moved (shifted and rotated) relative to the north pole, and each continent, in fact, shows its own polar wander path. During the late 1950s it was successfully shown on two occasions that these data could show the validity of continental drift: by Keith Runcorn in a paper in 1956 and by Warren Carey in a symposium held in March 1956.

The second piece of evidence in support of continental drift came during the late 1950s and early 60s from data on the bathymetry of the deep ocean floors and the nature of the oceanic crust such as magnetic properties and, more generally, with the development of marine geologywhich gave evidence for the association of seafloor spreading along the mid-oceanic ridges and magnetic field reversals, published between 1959 and 1963 by Heezen, Dietz, Hess, Mason, Vine & Matthews, and Morley.

Simultaneous advances in early seismic imaging techniques in and around Wadati-Benioff zones along the trenches bounding many continental margins, together with many other geophysical (e.g. gravimetric) and geological observations, showed how the oceanic crust could disappear into the mantle, providing the mechanism to balance the extension of the ocean basins with shortening along its margins.

All this evidence, both from the ocean floor and from the continental margins, made it clear around 1965 that continental drift was feasible and the theory of plate tectonics, which was defined in a series of papers between 1965 and 1967, was born, with all its extraordinary explanatory and predictive power. The theory revolutionized the Earth sciences, explaining a diverse range of geological phenomena and their implications in other studies such as paleogeography and paleobiology.

Continental Drift

In the late 19th and early 20th centuries, geologists assumed that the Earth's major features were fixed, and that most geologic features such as basin development and mountain ranges could be explained by vertical crustal movement, described in what is called the geosynclinal theory. Generally, this was placed in the context of a contracting planet Earth due to heat loss in the course of a relatively short geological time.

It was observed as early as 1596 that the opposite coasts of the Atlantic Ocean—or, more precisely, the edges of the continental shelves—have similar shapes and seem to have once fitted together.

Since that time many theories were proposed to explain this apparent complementarity, but the assumption of a solid Earth made these various proposals difficult to accept.

The discovery of radioactivity and its associated heating properties in 1895 prompted a re-examination of the apparent age of the Earth. This had previously been estimated by its cooling rate and assumption the Earth›s surface radiated like a black body. Those calculations had implied that, even if it started at red heat, the Earth would have dropped to its present temperature in a few tens of millions of years. Armed with the knowledge of a new heat source, scientists realized that the Earth would be much older, and that its core was still sufficiently hot to be liquid.

Alfred Wegener in Greenland in the winter of 1912-13.

By 1915, after having published a first article in 1912, Alfred Wegener was making serious arguments for the idea of continental drift in the first edition of *The Origin of Continents and Oceans*. In that book (re-issued in four successive editions up to the final one in 1936), he noted how the east coast of South America and the west coast of Africa looked as if they were once attached. Wegener was not the first to note this (Abraham Ortelius, Antonio Snider-Pellegrini, Eduard Suess, Roberto Mantovani and Frank Bursley Taylor preceded him just to mention a few), but he was the first to marshal significant fossil and paleo-topographical and climatological evidence to support this simple observation (and was supported in this by researchers such as Alex du Toit). Furthermore, when the rock strata of the margins of separate continents are very similar it suggests that these rocks were formed in the same way, implying that they were joined initially. For instance, parts of Scotland and Ireland contain rocks very similar to those found in Newfoundland and New Brunswick. Furthermore, the Caledonian Mountains of Europe and parts of the Appalachian Mountains of North America are very similar in structure and lithology.

However, his ideas were not taken seriously by many geologists, who pointed out that there was no apparent mechanism for continental drift. Specifically, they did not see how continental rock could plow through the much denser rock that makes up oceanic crust. Wegener could not explain the force that drove continental drift, and his vindication did not come until after his death in 1930.

Floating Continents, Paleomagnetism, and Seismicity Zones

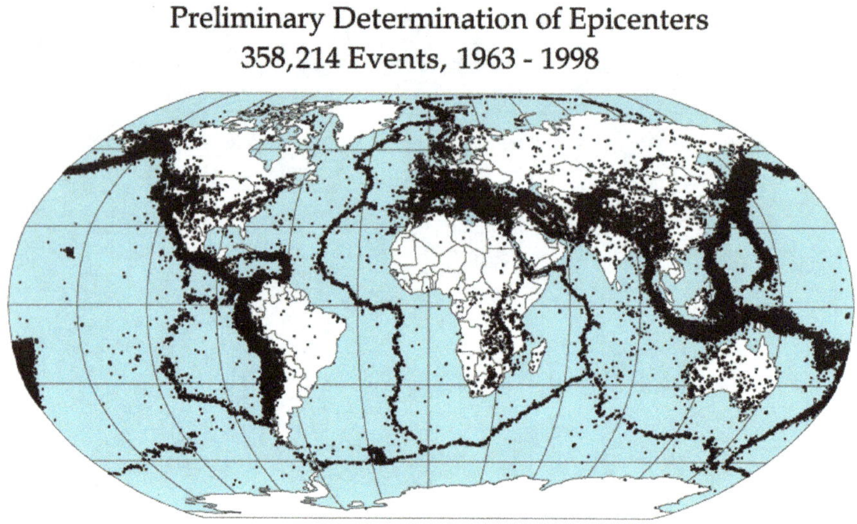

Global earthquake epicenters, 1963–1998

As it was observed early that although granite existed on continents, seafloor seemed to be composed of denser basalt, the prevailing concept during the first half of the twentieth century was that there were two types of crust, named "sial" (continental type crust) and "sima" (oceanic type crust). Furthermore, it was supposed that a static shell of strata was present under the continents. It therefore looked apparent that a layer of basalt (sial) underlies the continental rocks.

However, based on abnormalities in plumb line deflection by the Andes in Peru, Pierre Bouguer had deduced that less-dense mountains must have a downward projection into the denser layer underneath. The concept that mountains had "roots" was confirmed by George B. Airy a hundred years later, during study of Himalayan gravitation, and seismic studies detected corresponding density variations. Therefore, by the mid-1950s, the question remained unresolved as to whether mountain roots were clenched in surrounding basalt or were floating on it like an iceberg.

During the 20th century, improvements in and greater use of seismic instruments such as seismographs enabled scientists to learn that earthquakes tend to be concentrated in specific areas, most notably along the oceanic trenches and spreading ridges. By the late 1920s, seismologists were beginning to identify several prominent earthquake zones parallel to the trenches that typically were inclined 40–60° from the horizontal and extended several hundred kilometers into the Earth. These zones later became known as Wadati-Benioff zones, or simply Benioff zones, in honor of the seismologists who first recognized them, Kiyoo Wadati of Japan and Hugo Benioff of the United States. The study of global seismicity greatly advanced in the 1960s with the establishment of the Worldwide Standardized Seismograph Network (WWSSN) to monitor the compliance of the 1963 treaty banning above-ground testing of nuclear weapons. The much improved data from the WWSSN instruments allowed seismologists to map precisely the zones of earthquake concentration worldwide.

Meanwhile, debates developed around the phenomena of polar wander. Since the early debates of continental drift, scientists had discussed and used evidence that polar drift had occurred because continents seemed to have moved through different climatic zones during the past. Furthermore, paleomagnetic data had shown that the magnetic pole had also shifted during time. Reasoning in an opposite way, the continents might have shifted and rotated, while the pole remained relatively fixed. The first time the evidence of magnetic polar wander was used to support the movements of continents was in a paper by Keith Runcorn in 1956, and successive papers by him and his students Ted Irving (who was actually the first to be convinced of the fact that paleomagnetism supported continental drift) and Ken Creer.

This was immediately followed by a symposium in Tasmania in March 1956. In this symposium, the evidence was used in the theory of an expansion of the global crust. In this hypothesis the shifting of the continents can be simply explained by a large increase in size of the Earth since its formation. However, this was unsatisfactory because its supporters could offer no convincing mechanism to produce a significant expansion of the Earth. Certainly there is no evidence that the moon has expanded in the past 3 billion years; other work would soon show that the evidence was equally in support of continental drift on a globe with a stable radius.

During the thirties up to the late fifties, works by Vening-Meinesz, Holmes, Umbgrove, and numerous others outlined concepts that were close or nearly identical to modern plate tectonics theory. In particular, the English geologist Arthur Holmes proposed in 1920 that plate junctions might lie beneath the sea, and in 1928 that convection currents within the mantle might be the driving force. Often, these contributions are forgotten because:

- At the time, continental drift was not accepted.
- Some of these ideas were discussed in the context of abandoned fixistic ideas of a deforming globe without continental drift or an expanding Earth.
- They were published during an episode of extreme political and economic instability that hampered scientific communication.
- Many were published by European scientists and at first not mentioned or given little credit in the papers on sea floor spreading published by the American researchers in the 1960s.

Mid-oceanic Ridge Spreading and Convection

In 1947, a team of scientists led by Maurice Ewing utilizing the Woods Hole Oceanographic Institution›s research vessel *Atlantis* and an array of instruments, confirmed the existence of a rise in the central Atlantic Ocean, and found that the floor of the seabed beneath the layer of sediments consisted of basalt, not the granite which is the main constituent of continents. They also found that the oceanic crust was much thinner than continental crust. All these new findings raised important and intriguing questions.

The new data that had been collected on the ocean basins also showed particular characteristics regarding the bathymetry. One of the major outcomes of these datasets was that all along the globe, a system of mid-oceanic ridges was detected. An important conclusion was that along this system, new ocean floor was being created, which led to the concept of the Great Global Rift. This was described in the crucial paper of Bruce Heezen (1960), which would trigger a real revolution in thinking. A profound consequence of seafloor spreading is that new crust was, and still is, being continually created along the oceanic ridges. Therefore, Heezen advocated the so-called expanding Earth hypothesis of S. Warren Carey (see above). So, still the question remained: how can new crust be continuously added along the oceanic ridges without increasing the size of the Earth? In reality, this question had been solved already by numerous scientists during the forties and the fifties, like Arthur Holmes, Vening-Meinesz, Coates and many others: The crust in excess disappeared along what were called the oceanic trenches, where so-called subduction occurred. Therefore, when various scientists during the early sixties started to reason on the data at their disposal regarding the ocean floor, the pieces of the theory quickly fell into place.

The question particularly intrigued Harry Hammond Hess, a Princeton University geologist and a Naval Reserve Rear Admiral, and Robert S. Dietz, a scientist with the U.S. Coast and Geodetic Survey who first coined the term *seafloor spreading*. Dietz and Hess (the former published the same idea one year earlier in *Nature*, but priority belongs to Hess who had already distributed an unpublished manuscript of his 1962 article by 1960) were among the small handful who really understood the broad implications of sea floor spreading and how it would eventually agree with the, at that time, unconventional and unaccepted ideas of continental drift and the elegant and mobilistic models proposed by previous workers like Holmes.

In the same year, Robert R. Coats of the U.S. Geological Survey described the main features of island arc subduction in the Aleutian Islands. His paper, though little noted (and even ridiculed) at the time, has since been called seminal and prescient. In reality, it actually shows that the work by the European scientists on island arcs and mountain belts performed and published during the 1930s up until the 1950s was applied and appreciated also in the United States.

If the Earth›s crust was expanding along the oceanic ridges, Hess and Dietz reasoned like Holmes and others before them, it must be shrinking elsewhere. Hess followed Heezen, suggesting that new oceanic crust continuously spreads away from the ridges in a conveyor belt–like motion. And, using the mobilistic concepts developed before, he correctly concluded that many millions of years later, the oceanic crust eventually descends along the continental margins where oceanic trenches – very deep, narrow canyons – are formed, e.g. along the rim of the Pacific Ocean basin. The important step Hess made was that convection currents would be the driving force in this process, arriving at the same conclusions as Holmes had decades before with the only difference that the thinning of the ocean crust was performed using Heezen›s mechanism of spreading along the ridges. Hess therefore concluded that the Atlantic Ocean was expanding while the Pacific Ocean was shrinking. As old oceanic crust is "consumed" in the trenches (like Holmes and others, he thought this was done by thickening of the continental lithosphere, not, as now understood, by underthrusting at a larger scale of the oceanic crust itself into the mantle), new magma rises and erupts along the spreading ridges to form new crust. In effect, the ocean basins are perpetually being recycled, with the creation of new crust and the destruction of old oceanic lithosphere occurring simultaneously. Thus, the new mobilistic concepts neatly explained why the Earth does not get bigger with sea floor spreading, why there is so little sediment accumulation on the ocean floor, and why oceanic rocks are much younger than continental rocks.

Magnetic striping

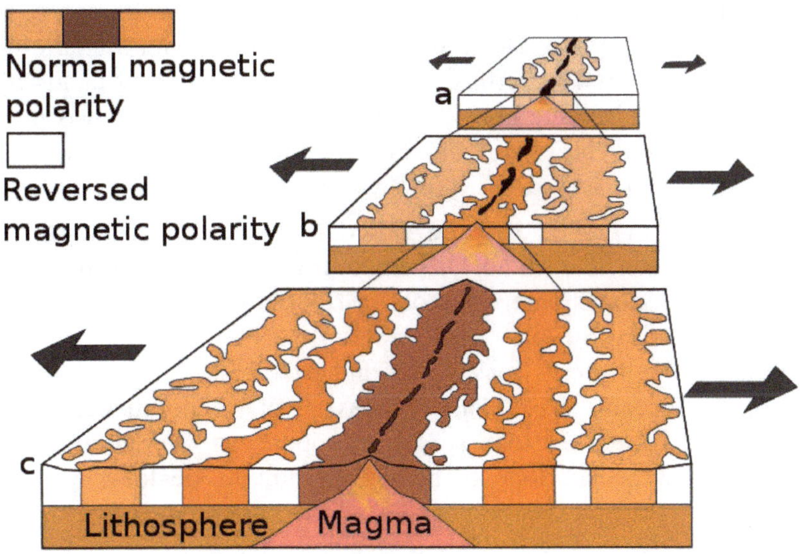

Seafloor magnetic striping.

A demonstration of magnetic striping. (The darker the color is, the closer it is to normal polarity)

Beginning in the 1950s, scientists like Victor Vacquier, using magnetic instruments (magnetometers) adapted from airborne devices developed during World War II to detect submarines, began recognizing odd magnetic variations across the ocean floor. This finding, though unexpected, was not entirely surprising because it was known that basalt—the iron-rich, volcanic rock making up the ocean floor—contains a strongly magnetic mineral (magnetite) and can locally distort compass readings. This distortion was recognized by Icelandic mariners as early as the late 18th century. More important, because the presence of magnetite gives the basalt measurable magnetic properties, these newly discovered magnetic variations provided another means to study the deep ocean floor. When newly formed rock cools, such magnetic materials recorded the Earth›s magnetic field at the time.

As more and more of the seafloor was mapped during the 1950s, the magnetic variations turned out not to be random or isolated occurrences, but instead revealed recognizable patterns. When these magnetic patterns were mapped over a wide region, the ocean floor showed a zebra-like pattern: one stripe with normal polarity and the adjoining stripe with reversed polarity. The overall pattern, defined by these alternating bands of normally and reversely polarized rock, became known as magnetic striping, and was published by Ron G. Mason and co-workers in 1961, who did not find, though, an explanation for these data in terms of sea floor spreading, like Vine, Matthews and Morley a few years later.

The discovery of magnetic striping called for an explanation. In the early 1960s scientists such as Heezen, Hess and Dietz had begun to theorise that mid-ocean ridges mark structurally weak zones where the ocean floor was being ripped in two lengthwise along the ridge crest (see the previous paragraph). New magma from deep within the Earth rises easily through these weak zones and eventually erupts along the crest of the ridges to create new oceanic crust. This process, at first denominated the "conveyer belt hypothesis and later called seafloor spreading, operating over many millions of years continues to form new ocean floor all across the 50,000 km-long system of mid-ocean ridges.

Only four years after the maps with the zebra pattern of magnetic stripes were published, the link between sea floor spreading and these patterns was correctly placed, independently by Lawrence Morley, and by Fred Vine and Drummond Matthews, in 1963, now called the Vine-Matthews-Morley hypothesis. This hypothesis linked these patterns to geomagnetic reversals and was supported by several lines of evidence:

- the stripes are symmetrical around the crests of the mid-ocean ridges; at or near the crest of the ridge, the rocks are very young, and they become progressively older away from the ridge crest;
- the youngest rocks at the ridge crest always have present-day (normal) polarity;
- stripes of rock parallel to the ridge crest alternate in magnetic polarity (normal-reversed-normal, etc.), suggesting that they were formed during different epochs documenting the (already known from independent studies) normal and reversal episodes of the Earth›s magnetic field.

By explaining both the zebra-like magnetic striping and the construction of the mid-ocean ridge system, the seafloor spreading hypothesis (SFS) quickly gained converts and represented another major advance in the development of the plate-tectonics theory. Furthermore, the oceanic crust now came to be appreciated as a natural tape recording of the history of the geomagnetic field reversals (GMFR) of the Earth›s magnetic field. Today, extensive studies are dedicated to the calibration of the normal-reversal patterns in the oceanic crust on one hand and known timescales derived from the dating of basalt layers in sedimentary sequences (magnetostratigraphy) on the other, to arrive at estimates of past spreading rates and plate reconstructions.

Definition and Refining of the Theory

After all these considerations, Plate Tectonics (or, as it was initially called "New Global Tectonics") became quickly accepted in the scientific world, and numerous papers followed that defined the concepts:

- In 1965, Tuzo Wilson who had been a promotor of the sea floor spreading hypothesis and continental drift from the very beginning added the concept of transform faults to the model, completing the classes of fault types necessary to make the mobility of the plates on the globe work out.
- A symposium on continental drift was held at the Royal Society of London in 1965 which must be regarded as the official start of the acceptance of plate tectonics by the scientific community, and which abstracts are issued as Blacket, Bullard & Runcorn (1965). In this symposium, Edward Bullard and co-workers showed with a computer calculation how the continents along both sides of the Atlantic would best fit to close the ocean, which became known as the famous "Bullard's Fit.
- In 1966 Wilson published the paper that referred to previous plate tectonic reconstructions, introducing the concept of what is now known as the Wilson Cycle.
- In 1967, at the American Geophysical Union›s meeting, W. Jason Morgan proposed that the Earth's surface consists of 12 rigid plates that move relative to each other.
- Two months later, Xavier Le Pichon published a complete model based on 6 major plates with their relative motions, which marked the final acceptance by the scientific community of plate tectonics.
- In the same year, McKenzie and Parker independently presented a model similar to Morgan's using translations and rotations on a sphere to define the plate motions.

Implications for biogeography

Continental drift theory helps biogeographers to explain the disjunct biogeographic distribution of present-day life found on different continents but having similar ancestors. In particular, it explains the Gondwanan distribution of ratites and the Antarctic flora.

PLATE RECONSTRUCTION

Reconstruction is used to establish past (and future) plate configurations, helping determine the shape and make-up of ancient supercontinents and providing a basis for paleogeography.

Defining Plate Boundaries

Current plate boundaries are defined by their seismicity. Past plate boundaries within existing plates are identified from a variety of evidence, such as the presence of ophiolites that are indicative of vanished oceans.

Past Plate Motions

Tectonic motion first began around three billion years ago.

Various types of quantitative and semi-quantitative information are available to constrain past plate motions. The geometric fit between continents, such as between west Africa and South America is still an important part of plate reconstruction. Magnetic stripe patterns provide a reliable guide to relative plate motions going back into the Jurassic period. The tracks of hotspots give absolute reconstructions, but these are only available back to the Cretaceous. Older reconstructions rely mainly on paleomagnetic pole data, although these only constrain the latitude and rotation, but not the longitude. Combining poles of different ages in a particular plate to produce apparent polar wander paths provides a method for comparing the motions of different plates through time. Additional evidence comes from the distribution of certain sedimentary rock types,faunal provinces shown by particular fossil groups, and the position of orogenic belts.

Formation and break-up of Continents

The movement of plates has caused the formation and break-up of continents over time, including occasional formation of a supercontinent that contains most or all of the continents. The supercontinent Columbia or Nuna formed during a period of 2,000 to 1,800 million years ago and broke up about 1,500 to 1,300 million years ago. The supercontinent Rodinia is thought to have formed about 1 billion years ago and to have embodied most or all of Earth's continents, and broken up into eight continents around 600 million years ago. The eight continents later re-assembled into another supercontinent called Pangaea; Pangaea broke up into Laurasia (which became North America and Eurasia) and Gondwana (which became the remaining continents).

The Himalayas, the world›s tallest mountain range, are assumed to have been formed by the collision of two major plates. Before uplift, they were covered by the Tethys Ocean.

CURRENT PLATES

There are dozens of smaller plates, the seven largest of which are the Arabian, Caribbean, Juan de Fuca, Cocos, Nazca, Philippine Sea and Scotia.

The current motion of the tectonic plates is today determined by remote sensing satellite data sets, calibrated with ground station measurements.

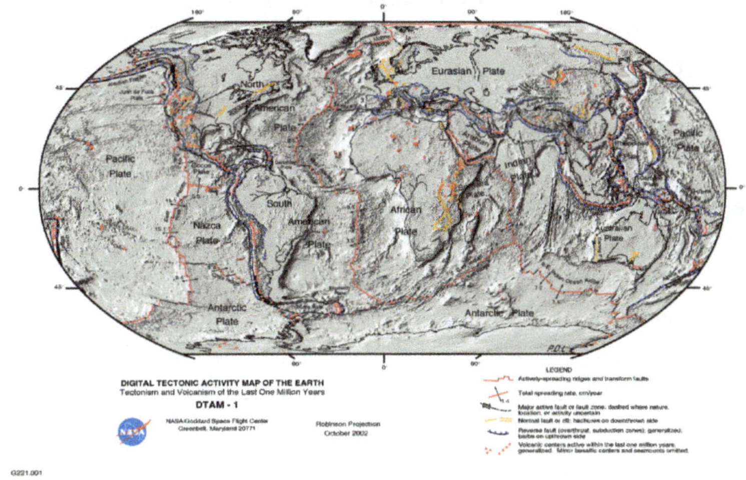

Depending on how they are defined, there are usually seven or eight "major" plates: African, Antarctic, Eurasian, North American, South American, Pacific, and Indo-Australian. The latter is sometimes subdivided into the Indian and Australian plates.

Other celestial bodies (planets, moons)

The appearance of plate tectonics on terrestrial planets is related to planetary mass, with more massive planets than Earth expected to exhibit plate tectonics. Earth may be a borderline case, owing its tectonic activity to abundant water (silica and water form a deep eutectic.)

Venus

Venus shows no evidence of active plate tectonics. There is debatable evidence of active tectonics in the planet's distant past; however, events taking place since then (such as the plausible and generally accepted hypothesis that the Venusian lithosphere has thickened greatly over the course of several hundred million years) has made constraining the course of its geologic record difficult. However, the numerous well-preserved impact craters have been utilized as a dating method to approximately date the Venusian surface (since there are thus far no known samples of Venusian rock to be dated by more reliable methods). Dates derived are dominantly in the range 500 to 750 million years ago, although ages of up to 1,200 million years ago have been calculated. This research has led to the fairly well accepted hypothesis that Venus has undergone an essentially complete volcanic resurfacing at least once in its distant past, with the last event taking place approximately within the range of estimated surface ages. While the mechanism of such an impressive thermal event remains a debated issue in Venusian geosciences, some scientists are advocates of processes involving plate motion to some extent.

One explanation for Venus›s lack of plate tectonics is that on Venus temperatures are too high for significant water to be present. The Earth›s crust is soaked with water, and water plays an important role in the development of shear zones. Plate tectonics requires weak surfaces in the crust along which crustal slices can move, and it may well be that such weakening never took place on Venus because of the absence of water. However, some researchers remain convinced that plate tectonics is or was once active on this planet.

Mars

Mars is considerably smaller than Earth and Venus, and there is evidence for ice on its surface and in its crust.

In the 1990s, it was proposed that Martian Crustal Dichotomy was created by plate tectonic processes. Scientists today disagree, and think that it was created either by upwelling within the Martian mantle that thickened the crust of the Southern Highlands and formed Tharsis or by a giant impact that excavated the Northern Lowlands.

Valles Marineris may be a tectonic boundary.

Observations made of the magnetic field of Mars by the *Mars Global Surveyor* spacecraft in 1999 showed patterns of magnetic striping discovered on this planet. Some scientists interpreted these as requiring plate tectonic processes, such as seafloor spreading. However, their data fail a “magnetic reversal test”, which is used to see if they were formed by flipping polarities of a global magnetic field.

Icy satellites

Some of the satellites of Jupiter have features that may be related to plate-tectonic style deformation, although the materials and specific mechanisms may be different from plate-tectonic activity on Earth. On 8 September 2014, NASA reported finding evidence of plate tectonics on Europa, a satellite of Jupiter—the first sign of such geological activity on another world other than Earth.

Titan, the largest moon of Saturn, was reported to show tectonic activity in images taken by the *Huygens* probe, which landed on Titan on January 14, 2005.

Exoplanets

On Earth-sized planets, plate tectonics is more likely if there are oceans of water; however, in 2007, two independent teams of researchers came to opposing conclusions about the likelihood of plate tectonics on larger super-earths with one team saying that plate tectonics would be episodic or stagnant and the other team saying that plate tectonics is very likely on super-earths even if the planet is dry.

REFERENCES

1. Little, Fowler & Coulson 1990.
2. Read & Watson 1975.
3. Scalera & Lavecchia 2006.
4. Zhen Shao 1997, Hancock, Skinner & Dineley 2000.
5. Turcotte & Schubert 2002, p. 5.

6. Turcotte & Schubert 2002.
7. Turcotte & Schubert 2002, p. 3.
8. Foulger 2010.
9. Schmidt & Harbert 1998.
10. Meissner 2002, p. 100.
11. Plate Tectonics: Plate Boundaries. platetectonics.com. Retrieved 12 June 2010.
12. Understanding plate motions. USGS. Retrieved 12 June 2010.
13. Grove, Timothy L.; Till, Christy B.; Krawczynski, Michael J. (8 March 2012). The Role of H2O in Subduction Zone Magmatism. Annual Review of Earth and Planetary Sciences (40): 413–439. Bibcode:2012AREPS..40..413G. doi:10.1146/annurev-earth-042711-105310. Retrieved 14 January 2016.
14. Mendia-Landa, Pedro. Myths and Legends on Natural Disasters: Making Sense of Our World. Retrieved 2008-02-05.
15. van Dijk 1992, van Dijk & Okkes 1991.
16. Holmes, Arthur (1931). Radioactivity and Earth Movements (PDF). Transactions of the Geological Society of Glasgow. Geological Society of Glasgow: 559–606.
17. Tanimoto & Lay 2000.
18. Meyerhoff et al. 1996.
19. Mallard, Claire; Coltice, Nicolas; Seton, Maria; Müller, R. Dietmar; Tackley, Paul J. (2016). Subduction controls the distribution and fragmentation of Earth›s tectonic plates. Nature. 535 *(7610): 140–143.* doi:10.1038/nature17992. ISSN 0028-0836.
20. Spence 1987, White & McKenzie 1989.
21. Conrad & Lithgow-Bertelloni 2002.
22. Spence 1987, White & Mckenzie 1989, Segev 2002.
23. Alfred Wegener (1880-1930). University of California Museum of Paleontology.
24. Neith, Katie (April 15, 2011). Caltech Researchers Use GPS Data to Model Effects of Tidal Loads on Earth›s Surface. Caltech. Retrieved August 15, 2012.
25. Ricard, Y. (2009). 2. Physics of Mantle Convection. In David Bercovici; Gerald Schubert. Treatise on Geophysics: Mantle Dynamics. 7. Elsevier Science. p. 36.
26. Glatzmaier, Gary A. (2013). Introduction to Modeling Convection in Planets and Stars: Magnetic Field, Density Stratification, Rotation. Princeton University Press. p. 149.
27. van Dijk 1992, van Dijk & Okkes 1990.
28. Moore 1973.
29. Bostrom 1971.
30. Scoppola et al. 2006.
31. Torsvik et al. 2010.
32. Wegener 1929.

33. Hughes 2001a.
34. Wegener 1966, Hughes 2001b.
35. Runcorn 1956.
36. Carey 1956.
37. see for example the milestone paper of Lyman & Fleming 1940.
38. Korgen 1995, Spiess & Kuperman 2003.
39. Kious & Tilling 1996.
40. Frankel 1987.
41. Joly 1909.
42. Thomson 1863.
43. Wegener 1912.
44. Stein & Wysession 2009, p. 26
45. Carey 1956; see also Quilty 2003.
46. Holmes 1928; see also Holmes 1978, Frankel 1978.
47. Lippsett 2001, Lippsett 2006.
48. Heezen 1960.
49. Dietz 1961.
50. Hess 1962.
51. Mason & Raff 1961, Raff & Mason 1961.
52. Vine & Matthews 1963.
53. See summary in Heirzler, Le Pichon & Baron 1966
54. Wilson 1963.
55. Wilson 1965.
56. Wilson 1966.
57. Morgan 1968.
58. Le Pichon 1967.
59. McKenzie & Parker 1967.
60. Moss & Wilson 1998.
61. Condie 1997.
62. Lliboutry 2000.
63. Kranendonk, V.; Martin, J. (2011). Onset of Plate Tectonics. Science. 333 *(6041): 413–414. Bibcode*:2011Sci...333..413V. doi:10.1126/science.1208766. PMID 21778389.
64. Torsvik, Trond Helge. Reconstruction Methods. Retrieved 18 June 2010.
65. Torsvik 2008.
66. Butler 1992.
67. Scotese, C.R. (2002-04-20). Climate History. Paleomap Project. Retrieved 18

June 2010.

68. Zhao 2002, 2004

69. Valencia, O›Connell & Sasselov 2007.

70. Kasting 1988.

71. Bortman, Henry (2004-08-26). Was Venus alive? The Signs are Probably There. Astrobiology Magazine. Retrieved 2008-01-08.

72. Sleep 1994.

73. Zhong & Zuber 2001.

74. Andrews-Hanna, Zuber & Banerdt 2008.

75. Wolpert, Stuart (August 9, 2012). UCLA scientist discovers plate tectonics on Mars. Yin, An. UCLA. Retrieved August 13, 2012.

76. Connerney et al. 1999, Connerney et al. 2005

77. Harrison 2000.

78. Dyches, Preston; Brown, Dwayne; Buckley, Michael (8 September 2014). Scientists Find Evidence of ‹Diving› Tectonic Plates on Europa. NASA. Retrieved 8 September 2014.

79. Soderblom et al. 2007.

80. Valencia, Diana; O›Connell, Richard J. Convection scaling and subduction on Earth and super-Earths. Earth and Planetary Science Letters. 286 *(3–4): 492–502. Bibcode*:2009E&PSL.286..492V. doi:10.1016/j.epsl.2009.07.015.

81. van Heck, H. J.; Tackley, P.J. Plate tectonics on super-Earths: Equally or more likely than on Earth. Earth and Planetary Science Letters. 310 *(3–4): 252–261. Bibcode*:2011E&PSL.310..252V. doi:10.1016/j.epsl.2011.07.029.

82. C. O'Neill, A. Lenardic Geological consequences of super-sized Earths GEOPHYSICAL RESEARCH LETTERS 34: L19204 doi:10.1029/2007GL030598

CHAPTER 6

Structural Geology

Chapter Contents

- **Introduction**
- **Use and Importance**
- **Methods**

Structural geology is a subfield within geology which focuses on the study of geological structures, with the goal of learning how, when, and why they formed. There are a number of applications for structural geology, ranging from determining where valuable mineral resources might be buried to assessing land to determine whether or not it is safe to build on. Practitioners in this field usually have a bachelor's degree in structural geology, and they may have pursued graduate work as well.

INTRODUCTION

Structural geology is the study of the three-dimensional distribution of rock units with respect to their deformational histories. The primary goal of structural geology is to use measurements of present-day rock geometries to uncover information about the history of deformation (strain) in the rocks, and ultimately, to understand the stress field that resulted in the observed strain and geometries. This understanding of the dynamics of the stress field can be linked to important events in the geologic past; a common goal is to understand the structural evolution of a particular area with respect to regionally widespread patterns of rock deformation (e.g., mountain building, rifting) due to plate tectonics.

USE AND IMPORTANCE

The study of geologic structures has been of prime importance in economic geology, both petroleum geology and mining geology. Folded and faulted rock strata commonly form traps that accumulate and concentrate fluids such as petroleum and natural gas. Similarly, faulted and structurally complex areas are notable as permeable zones for hydrothermal fluids, resulting in concentrated areas of base and precious metal ore deposits. Veins of minerals containing various metals commonly occupy faults and fractures in structurally complex areas. These structurally fractured and faulted zones often occur in association with intrusive igneous rocks. They often also occur around geologic reef complexes and collapse features such as ancient sinkholes. Deposits of gold, silver, copper, lead, zinc, and other metals, are commonly located in structurally complex areas.

Structural geology is a critical part of engineering geology, which is concerned with the physical and mechanical properties of natural rocks. Structural fabrics and defects such as faults, folds, foliations and joints are internal weaknesses of rocks which may affect the stability of human engineered structures such as dams, road cuts, open pit mines and underground mines or road tunnels.

Geotechnical risk, including earthquake risk can only be investigated by inspecting a combination of structural geology and geomorphology. In addition, areas of karst landscapes which reside atop underground caverns, potential sinkholes, or other collapse features are of particular importance for these scientists. In addition, areas of steep slopes are potential collapse or landslide hazards. Environmental geologists and hydrogeologists need to apply the tenets of structural geology to understand how geologic sites impact (or are impacted by) groundwater flow and penetration. For instance, a hydrogeologist may need to determine if seepage of toxic substances from waste dumps is occurring in a residential area or if salty water is seeping into an aquifer.

Plate tectonics is a theory developed during the 1960s which describes the movement of continents by way of the separation and collision of crustal plates. It is in a sense structural geology on a planet scale, and is used throughout structural geology as a framework to analyze and understand global, regional, and local scale features.

METHODS

Structural geologists use a variety of methods to (first) measure rock geometries, (second) reconstruct their deformational histories, and (third) calculate the stress field that resulted in that deformation.

Geometries

Primary data sets for structural geology are collected in the field. Structural geologists measure a variety of planar features (bedding planes, foliation planes, fold axial planes, fault planes, and joints), and linear features (stretching lineations, in which minerals are ductily extended; fold axes; and intersection lineations, the trace of a planar feature on another planar surface).

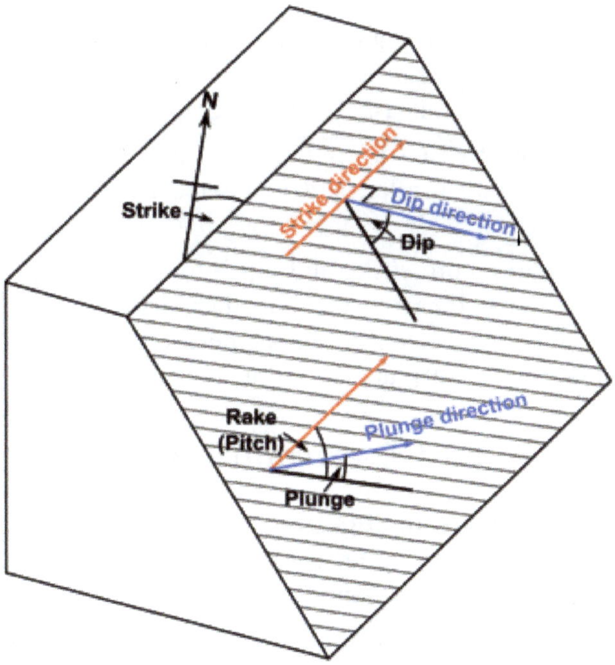

Illustration of measurement conventions for planar and linear structures

Measurement Conventions

The inclination of a planar structure in geology is measured by *strike and dip*. The strike is the line of intersection between the planar feature and a horizontal plane, taken according to the right hand convention, and the dip is the magnitude of the inclination, below horizontal, at right angles to strike. For example; striking 25 degrees East of North, dipping 45 degrees Southeast, recorded as N25E,45SE.

Alternatively, dip and dip direction may be used as this is absolute. Dip direction is measured in 360 degrees, generally clockwise from North. For example, a dip of 45 degrees towards 115 degrees azimuth, recorded as 45/115. Note that this is the same as above.

The term *hade* is occasionally used and is the deviation of a plane from vertical i.e. (90°-dip).

Fold axis plunge is measured in dip and dip direction (strictly, plunge and azimuth of plunge). The orientation of a fold axial plane is measured in strike and dip or dip and dip direction.

Lineations are measured in terms of dip and dip direction, if possible. Often lineations occur expressed on a planar surface and can be difficult to measure directly. In this case, the lineation may be measured from the horizontal as a *rake* or *pitch* upon the surface.

Rake is measured by placing a protractor flat on the planar surface, with the flat edge horizontal and measuring the angle of the lineation clockwise from horizontal. The orientation of the lineation can then be calculated from the rake and strike-dip information of the plane it was measured from, using a stereographic projection.

If a fault has lineations formed by movement on the plane, e.g.; slickensides, this is recorded as a lineation, with a rake, and annotated as to the indication of throw on the fault.

Generally it is easier to record strike and dip information of planar structures in dip/dip direction format as this will match all the other structural information you may be recording about folds, lineations, etc., although there is an advantage to using different formats that discriminate between planar and linear data.

Plane, Fabric, Fold and Deformation Conventions

The convention for analysing structural geology is to identify the planar structures, often called *planar fabrics* because this implies a textural formation, the linear structures and, from analysis of these, unravel deformations.

Planar structures are named according to their order of formation, with original sedimentary layering the lowest at S0. Often it is impossible to identify S0 in highly deformed rocks, so numbering may be started at an arbitrary number or given a letter (S_A, for instance). In cases where there is a bedding-plane foliation caused by burial metamorphism or diagenesis this may be enumerated as S0a.

If there are folds, these are numbered as F_1, F_2, etc. Generally the axial plane foliation or cleavage of a fold is created during folding, and the number convention should match. For example, an F_2 fold should have an S_2 axial foliation.

Deformations are numbered according to their order of formation with the letter D denoting a deformation event. For example, D_1, D_2, D_3. Folds and foliations, because they are formed by deformation events, should correlate with these events. For example, an F_2 fold, with an S_2 axial plane foliation would be the result of a D_2 deformation.

Metamorphic events may span multiple deformations. Sometimes it is useful to identify them similarly to the structural features for which they are responsible, e.g.; M_2. This may be possible by observing porphyroblast formation in cleavages of known deformation age, by identifying metamorphic mineral assemblages created by different events, or via geochronology.

Intersection lineations in rocks, as they are the product of the intersection of two planar structures, are named according to the two planar structures from which they are formed. For instance, the intersection lineation of a S_1 cleavage and bedding is the L_{1-0} intersection lineation (also known as the cleavage-bedding lineation).

Stretching lineations may be difficult to quantify, especially in highly stretched ductile rocks where minimal foliation information is preserved. Where possible, when correlated with deformations (as few are formed in folds, and many are not strictly associated with planar foliations), they may be identified similar to planar surfaces and folds, e.g.; L_1, L_2.

For convenience some geologists prefer to annotate them with a subscript S, for example L_{s1} to differentiate them from intersection lineations, though this is generally redundant.

Stereographic Projections

Stereographic projection of structural strike and dip measurements is a powerful method for analyzing the nature and orientation of deformation stresses, lithological units and penetrative fabrics.

Rock macro-Structures

On a large scale, structural geology is the study of the three-dimensional interaction and relationships of stratigraphic units within terranes of rock or geological regions.

This branch of structural geology deals mainly with the orientation, deformation and relationships of stratigraphy (bedding), which may have been faulted, folded or given a foliation by some tectonic event. This is mainly a geometric science, from which *cross sections* and three-dimensional *block models* of rocks, regions, terranes and parts of the Earth's crust can be generated.

Study of regional structure is important in understanding orogeny, plate tectonics and more specifically in the oil, gas and mineral exploration industries as structures such as faults, folds and unconformities are primary controls on ore mineralisation and oil traps.

Modern regional structure is being investigated using seismic tomography and seismic reflection in three dimensions, providing unrivaled images of the Earth's interior, its faults and the deep crust. Further information from geophysics such as gravity and airborne magnetics can provide information on the nature of rocks imaged to be in the deep crust.

Rock Microstructures

Rock microstructure includes the texture of a rock and the small scale rock structures. The words "texture" and "microstructure" are interchangeable, with the latter preferred in modern geological literature. However, texture is still acceptable because it is a useful means of identifying the origin of rocks, how they formed, and their appearance.

Textures are *penetrative fabrics* of rocks; they occur throughout the entirety of the rock mass on a microscopic, hand specimen and often on an outcrop scale. This is similar in many ways to foliations, except a texture does not necessarily carry structural information in terms of deformation events and orientation information. Structures occur on hand-specimen scale and above.

Microstructure analysis describes the textural features of the rock, and can provide information on the conditions of formation, petrogenesis, and subsequent deformation, folding or alteration events.

Sedimentary Microstructures

Description of sedimentary rock microstructure aims to provide information on the conditions of deposition of the sediment, the paleo-environment, and the provenance of the sedimentary material.

Methods involve description of clast size, sorting, composition, rounding or angularity, sphericity and description of the matrix. Sedimentary microstructures, specifically, may include microscopic analogs of larger sedimentary structural features such as cross-bedding, syn-sedimentary faults, sediment slumping, cross-stratification, etc.

Maturity

The maturity of a sediment is related not only to the sorting (mean grain size and deviations), but also to the fragment sphericity, rounding and composition. Quartz-only sands are more mature than arkose or greywacke.

Fragment shape

Fragment shape gives information on the length of sediment transport. The more rounded the clasts, the more water-worn they are. Particle shape includes form and rounding. Form indicates whether a grain is more equant (round, spherical) or platy (flat, disc-like, oblate); as well as sphericity.

Roundness

Roundness refers to the degree of sharpness of the corners and edges of a grain. The surface texture of grains may be polished, frosted, or marked by small pits and scratches. This information can usually be seen best under a binocular microscope, not in a thin section.

Composition

Composition of the clasts can give clues as to the derivation of a rock's sediments. For instance, volcanic fragments, fragments of cherts, well-rounded sands all imply different sources.

Matrix and cement

The matrix of a sedimentary rock and the mineral cement (if any) holding it together are all diagnostic.

Diagenetic features

Usually diagenesis results in a weak bedding-plane foliation. Other effects can include flattening of grains, pressure dissolution and sub-grain deformation. Mineralogical changes may include zeolite or other authigenic minerals forming in low-grade metamorphic conditions.

Sorting

Sorting is used to describe the uniformity of grain sizes within a sedimentary rock. Understanding sorting is critical to making inferences on the degree of maturity and length of transport of a sediment. Sediments become sorted on the basis of density, because of the energy of the transporting medium. High energy currents can carry larger fragments. As the energy decreases, heavier particles are deposited and lighter fragments continue to be transported. This results in sorting due to density. Sorting can be expressed mathematically by the standard deviation of the grain-size frequency curve of a sediment sample, expressed as values of φ (phi). Values range from <0.35φ (very well sorted) to >4.00φ (extremely poorly sorted).

Metamorphic microstructure

The study of metamorphic rock microstructures aims to determine the timing, sequence and conditions of deformations, mineral growth and overprinting of subsequent deformation events. Metamorphic microstructures include textures formed by the development of foliation and overprinting of foliations causing crenulations. The relationship of porphyroblasts to the foliations and to other porphyroblasts can provide information on the order of formation of metamorphic assemblages or facies of minerals.

Shear textures are particularly suited to analysis by microstructural investigations, especially in mylonites and other highly disturbed and deformed rocks.

Foliations and crenulations

On the thin section and hand specimen scale a metamorphic rock may manifest a planar penetrative fabric called a foliation or a cleavage. Several foliations may be present in a rock, giving rise to a crenulation. Identifying a foliation and its orientation is the first step in analysis of foliated metamorphic rocks. Gaining information on when the foliation formed is essential to reconstructing a P-T-t (pressure, temperature, time) path for a rock, as the relationship of a foliation to porphyroblasts is diagnostic of when the foliation formed, and the P-T conditions which existed at that time.

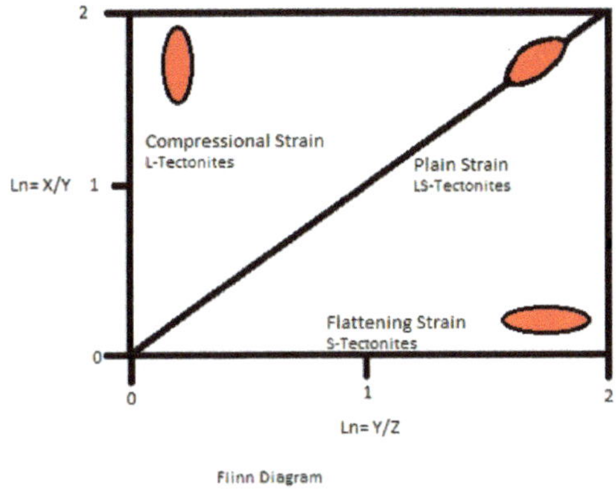

Flinn Diagram showing degree of stretching, or lineation (L) versus flattening, or foliation (S)

Lineations

Linear structures in a rock may arise from the intersection of two foliations or planar structures, such as a sedimentary bedding plane and a tectonically induced cleavage plane. The degree of lineation compared with the degree of foliation for certain strain markers in deformed rocks are commonly plotted on a Flinn diagram.

Ductile Shear Microstructures

Very distinctive textures form as a consequence of ductile shear. The microstructures of ductile shear zones are S-planes, C-planes and C' planes. S-planes or *schistosity* planes are parallel with the shear direction and are generally defined by micas or platy minerals. Define the flattened long-axis of the strain ellipse. C-planes or *cissalement* planes form oblique to the shear plane. The angle between the C and S planes is always acute, and defines the shear sense. Generally, the lower the C-S angle the greater the strain. The C' planes are rarely observed except in ultradeformed mylonites, and form nearly perpendicular to the S-plane.

Other microstructures which can give sense of shear include

- sigmoidal veins
- mica fish
- rotated porphyroblasts

Igneous Microstructure

Analysis of igneous rock microstructure may complement descriptions on the hand specimen and outcrop scale. This is especially vital for describing phenocrysts and fragmental textures of tuffs, as often relationships between magma and phenocryst morphology are critical for analysing cooling, fractional crystallization and emplacement.

Analysis of intrusive rock microstructures can provide information on source and genesis, including contamination of igneous rocks by wall rocks and identifying crystals which may have been accumulated or dropped out of the melt. This is especially critical for komatiite lavas and ultramafic intrusive rocks.

General principles of Igneous Microstructure

Igneous microstructure is a combination of cooling rate, nucleation rate, eruption (if a lava), magma composition and its relationships to what minerals will nucleate, as well as physical effects of wall rocks, contamination and especially vapor.

Grain Texture

According to the texture of the grains, igneous rocks may be classified as

- pegmatitic: very large crystals

- phaneritic: rocks contain minerals with crystals visible to the unaided eye, commonly intrusive
- aphanitic: rapid cooling, crystal nucleation and growth is stunted, forming a uniform, fine grained rock
- porphyritic: containing phenocrysts in a fine groundmass
- vesicular: contains voids caused by trapped gas while cooling
- vitreous: glassy or hyaline without crystals
- pyroclastic: rock contain fragments of crystals, phenocrysts and rock fragments
- equigranular: rock crystals are all the same size

Crystal Shapes

Crystal shape is also an important factor in the texture of an igneous rock. Crystals may be euhedral, subeuhedral or anhedral:

- *Euhedral* or *automorphic*, if the crystallographic shape is preserved.
- *Subeuhedral* or *Subhedral*, if only part is preserved.
- *Anhedral* or xenomorphic, if the crystals present no recognizable crystallographic forms.

Rocks composed entirely of euhedral crystals are termed *panidiomorphic*, and rocks composed entirely of subhedral crystals are termed *subidiomorphic*.

Porphyritic Structure

Porphyritic structure is caused by the nucleation of crystal sites and the growth of crystals in a liquid magma. Often a magma can only grow one mineral at a time especially if it is cooling slowly. This is why most igneous rocks have only one type of phenocryst mineral. Rhythmic cumulate layers in ultramafic intrusions are a result of uninterrupted slow cooling.

When a rock cools too quickly the liquid freezes into a solid glass, or crystalline groundmass. Often vapor loss from a magma chamber will cause a porphyritic texture.

Embayments or 'corroded' margins to phenocrysts infer that they were being resorbed by the magma and may imply addition of fresh, hotter magma. Ostwald ripening is also used to explain some porphyritic igneous textures, especially orthoclase megacrystic granites.

Phenocryst Shape: Implications

A crystal growing in a magma adopts a habit (see crystallography) which best reflects its environment and cooling rate. The usual phenocryst habit is the ones commonly observed. This may imply a 'normal' cooling rate.

Abnormal cooling rates occur in supercooled magmas, particularly komatiite lavas. Here, low nucleation rates due to superfluidity prevent nucleation until the liquid is well below the mineral growth curve. Growth then occurs at extreme rates, favoring slender, long crystals. Additionally, at crystal vertices and terminations, spikes and skeletal shapes may form because nucleation favors crystal edges. Spinifex or dendritic texture is an example of this result. Hence, the shape of phenocrysts can provide valuable information on cooling rate and initial magma temperature.

Spherulites

Spherulitic texture is the result of cooling and nucleation of material in a magma which has achieved supersaturation in the crystal component. Thus it is often a subsolidus process in supercooler felsic rocks. Often, two minerals will grow together in the spherulite. Axiolitic texture results from spherulitic growth along fractures in volcanic glass, often from invasion of water.

Graphic and other Intergrowth Textures

Intergrowths of two or more minerals can form in a variety of ways, and interpretations of the intergrowths can be critical in understanding both magmatic and cooling histories of igneous rocks. A few of the many important textures are presented here as examples.

Graphic, micrographic texture, and granophyric textures are examples of intergrowths formed during magmatic crystallization. They are angular intergrowths of quartz and alkali feldspar. When well-developed, the intergrowths may resemble ancient cuneiform writing, hence the name. These intergrowths are typical of pegmatite and granophyre, and they have been interpreted as documenting simultaneous crystallization of the intergrown minerals in the presence of a silicate melt together with a water-rich phase.

Intergrowths that form by exsolution are aids in interpreting cooling histories of rocks. Perthite is an intergrowth of K-feldspar with albite feldspar, formed by exsolution from an alkali feldspar of intermediate composition: the coarseness of perthitic intergrowths is related to cooling rate. Perthite is typical of many granites. Myrmekite is a microscopic, vermicular (worm-like) intergrowth of quartz and sodium-rich plagioclase common in granite; myrmekite may form as alkali feldspar breaks down by exsolution and silicon is transported by fluids in cooling rocks.

Iron-titanium oxides are extremely important, as they carry the predominant magnetic signatures of many rocks, and so they have played a major role in our understanding of plate tectonics. These oxides commonly have complex textures related both to exsolution and oxidation. For instance, ulvospinel in igneous rocks such as basalt and gabbro commonly oxidizes during subsolidus cooling to produce regular intergrowths of magnetite and ilmenite. The process can determine what magnetic record is inherited by the rock.

Kinematics

Geologists use rock geometry measurements to understand the history of strain in rocks. Strain can take the form of brittle faulting and ductile folding and shearing. Brittle deformation takes place in the shallow crust, and ductile deformation takes place in the deeper crust, where temperatures and pressures are higher.

Stress Fields

By understanding the constitutive relationships between stress and strain in rocks, geologists can translate the observed patterns of rock deformation into a stress field during the geologic past. The following list of features are typically used to determine stress fields from deformational structures.

- In perfectly brittle rocks, faulting occurs at 30° to the greatest compressional stress. (Byerlee's Law)
- The greatest compressive stress is normal to fold axial planes.

REFERENCES

1. Russell, William L (1955). "1. Introduction". Structural Geology for Petroleum Geologists. New York: McGraw-Hill. p. 1.
2. "Plate tectonics and people". USGS.
3. Livaccari, Richard F.; Burke, Kevin; Scedilengör, A. M. C. "Was the Laramide orogeny related to subduction of an oceanic plateau?". *Nature.* 289 *(5795): 276–278. Bibcode*:1981Natur.289..276L. doi:10.1038/289276a0.
4. M. King Hubbert (1972). Structural Geology. Hafner Publishing Company.
5. G.H. Davis and S.J. Reynolds (1996). The structural geology of rocks and regions (2nd ed.). Wiley. ISBN 0-471-52621-5.
6. C.W. Passchier and R.A.J. Trouw (1998). Microtectonics. Berlin: Springer. ISBN 3-540-58713-6.
7. B.A. van der Pluijm and S. Marshak (2004). Earth Structure - An Introduction to Structural Geology and Tectonics (2nd ed.). New York: W. W. Norton. p. 656. ISBN 0-393-92467-X.
8. V. Guerriero; et al. (2011). "Improved statistical multi-scale analysis of fractures in carbonate reservoir analogues". Tectonophysics. Elsevier. 504: 14–24. Bibcode:2011Tectp.504...14G. doi:10.1016/j.tecto.2011.01.003.
9. V. Guerriero; et al. (2009). "Quantifying uncertainties in multi-scale studies of fractured reservoir analogues: Implemented statistical analysis of scan line data from carbonate rocks". Journal of Structural Geology. Elsevier. 32 *(9): 1271–1278. Bibcode*:2010JSG....32.1271G. doi:10.1016/j.jsg.2009.04.016.

CHAPTER 7

Planetary Geology and Geology of Solar Terrestrial Planets

Chapter Contents

Terrestrial planets are Earth-like planets (in Latin, terra means Earth) made up of rocks or metals with a hard surface — making them different from other planets that lack a solid surface. Terrestrial planets also have a molten heavy metal core, few moons, and topological features such as valleys, volcanoes and craters.

In our solar system, there are four terrestrial planets, which also happen to be the four closest to the sun: Mercury, Venus, Earth and Mars. During the creation of the solar system, there were likely more terrestrial planetoids, but they likely merged or were destroyed.

PLANETARY GEOLOGY

Planetary geology, alternatively known as **astrogeology** or **exogeology**, is a planetary science discipline concerned with the geology of the celestial bodies such as the planets and their moons, asteroids, comets, and meteorites. Although the geo- prefix typically indicates topics of or relating to the Earth, planetary geology is named as such for historical and convenience reasons; applying geological science to other planetary bodies. Due to the types of investigations involved, it is also closely linked with Earth-based geology.

Planetary geologist and NASAastronaut Harrison Jack Schmitt collecting lunar samples during the Apollo 17 mission in early-December 1972

Surface of Mars as photographed by the Viking 2 lander December 9, 1977.

Planetary geology includes such topics as determining the internal structure of the terrestrial planets, and also looks at planetary volcanism and surface processes such as impact craters, fluvial and aeolian processes. The structures of the giant planets and their moons are also examined, as is the make-up of the minor bodies of the Solar System, such as asteroids, the Kuiper Belt, and comets.

HISTORY OF PLANETARY GEOLOGY

Eugene Shoemaker is credited with bringing geologic principles to planetary mapping and creating the branch of planetary science in the early 1960s, the Astrogeology Research Program, within the United States Geological Survey. He made important contributions to the field and the study of impact craters, Selenography (study of the Moon), asteroids, and comets. Today many institutions are concerned with the study and communication of planetary sciences and planetary geology. The Visitor Center at Barringer Meteor Crater near Winslow, Arizona includes a Museum of planetary geology. The Geological Society of America's Planetary Geology Division has been growing and thriving since May 1981 and has two mottos: "One planet just isn't enough!" and ""“The GSA Division with the biggest field area!" Major centers for planetary science research include the Lunar and Planetary Institute, the Applied Physics Laboratory, the Planetary Science Institute, the Jet Propulsion Laboratory, Southwest Research Institute, and Johnson Space Center. Additionally, several universities conduct extensive planetary science research, including Montana State University, Brown University, the University of Arizona, Caltech, the University of Colorado, Western Michigan University, MIT, and Washington University in St. Louis.

FEATURES AND TERMS

Planetary geology uses a wide variety of standardised descriptor names for features. All planetary feature names recognised by the International Astronomical Union combine one of these names with a possibly unique identifying name. The conventions which decide the more precise name are dependent on which planetary body the feature is on, but the standard descriptors are in general common to all astronomical planetary bodies. Some names have a long history of historical usage, but new must be recognised by the IAU Working Group for Planetary System Nomenclature as features are mapped and described by new planetary missions. This means that in some cases names may change as new imagery becomes available, or in other cases widely adopted informal names changed in line with the rules. The standard names are chosen to consciously avoid interpreting the underlying cause of the feature, but rather to describe only its appearance.

Feature	Pronunciation[7]	Description	Designation
Albedo feature	/æl'biːdoʊ/	An area which shows a contrast in brightness or darkness (albedo) with adjacent areas. This term is implicit.	AL
Arcus, arcūs	/'ɑːrkəs/	Arc: curved feature	AR
Astrum, astra	/'æstrəm/, /'æstrə/	Radial-patterned features on Venus	AS
Catena, catenae	/kə'tiːnə/, /kə'tiːniː/	A chain of craters e.g. Enki Catena.	CA
Cavus, cavi	/'keɪvəs/, /'keɪvaɪ/	Hollows, irregular steep-sided depressions usually in arrays or clusters	CB
Chaos	/'keɪ.ɒs/	A distinctive area of broken or jumbled terrain e.g. Iani Chaos.	CH
Chasma, chasmata	/'kæzmə/, /'kæzmətə/	Deep, elongated, steep-sided depression e.g. Eos Chasma.	CM
Colles	/'kɒliːz/	A collection of small hills or knobs.	CO

Corona, coronae	/kɒ'roʊnə/, /kɒ'roʊniː/	An oval feature. Used only on Venus and Miranda.	CR
Crater, craters	/'kreɪtər/	A circular depression likely created by impact event. This term is implicit.	AA
Dorsum, dorsa	/'dɔːrsəm/, /'dɔːrsə/	Ridge, sometimes called a wrinkle ridge e.g. Dorsum Buckland.	DO
Eruptive center		An active volcano on Io. This term is implicit.	ER
Facula, faculae	/'fækjʊlə/, /'fækjʊliː/	Bright spot	FA
Farrum, farra	/'færəm/, /'færə/	Pancake-like structure, or a row of such structures. Used only on Venus.	FR
Flexus, flexūs	/'flɛksəs/	Very low curvilinear ridge with a scalloped pattern	FE
Fluctus, fluctūs	/'flʌktəs/	Terrain covered by outflow of liquid. Used on Venus, Io and Titan.	FL

Flumen, flumina	/'fluːm□n/, /'fluːm□nə/	Channel on Titan that might carry liquid	FM
Fossa, fossae	/'fɒsə/, /'fɒsiː/	Long, narrow, shallow depression	FO
Fretum, freta	/'friːtəm/, /'friːtə/	Strait of liquid connecting two larger areas of liquid. Used only on Titan.	FT
Insula, insulae	/'ɪnsjuːlə/, /'ɪnsjuːliː/	Island (islands), an isolated land area (or group of such areas) surrounded by, or nearly surrounded by, a liquid area (sea or lake). Used only on Titan.	IN
Labes, labes	/'leɪbiːz/	Landslide debris. Used only on Mars.	LA

Labyrinthus, labyrinthi	/læb□'rɪnθəs/, /læb□'rɪnθaɪ/	Complex of intersecting valleys or ridges.	LB
Lacuna, lacunae	/lə'kjuːnə/, /lə'kjuːniː/	Irregularly shaped depression having the appearance of a dry lake bed. Used only on Titan.	LU
Lacus, lacūs	/'leɪkəs/	A "lake" or small plain on Moon and Mars; on Titan, a "lake" or small, dark plain with discrete, sharp boundaries.	LC
Landing site name		Lunar features at or near Apollo landing sites	LF
Large ringed feature		Cryptic ringed features	LG
Lenticula, lenticulae	/lɛn'tɪkjʊlə/, /lɛn'tɪkjʊliː/	Small dark spots on Europa	LE
Linea, lineae	/'lɪniːə/, /'lɪniː.iː/	Dark or bright elongate marking, may be curved or straight	LI
Maçula, maculae	/'mækjʊlə/, /'mækjʊliː/	Dark spot, may be irregular	MA

Mare, maria	/ˈmɑːriː/ ~ /ˈmɑːreɪ/, /ˈmɑːriə/	A "sea" or large circular plain on Moon and Mars, e.g. Mare Erythraeum; on Titan, large expanses of dark materials thought to be liquid hydrocarbons, e.g. Ligeia Mare.	ME
Mensa, mensae	/ˈmɛnsə/, /ˈmɛnsiː/	A flat-topped prominence with cliff-like edges, i.e. a mesa.	MN
Mons, montes	/ˈmɒnz/, /ˈmɒntiːz/	Mons refers to a mountain. Montes refers to a mountain range.	MO
Oceanus	/oʊʃiːˈeɪnəs/	Very large dark area. The only feature with this designation is Oceanus Procellarum.	OC
Palus, paludes	/ˈpeɪləs/, /pəˈljuːdiːz/	"Swamp"; small plain. Used on the Moon and Mars.	PA
Patera, paterae	/ˈpætərə/, /ˈpætəriː/	Irregular crater, or a complex one with scalloped edges e.g. Ah Peku Patera. Usually refers to the dish-shaped depression atop a volcano.	PE
Planitia, planitiae	/pləˈnɪʃə/, /pləˈnɪʃiː/	Low plain e.g. Amazonis Planitia.	PL
Planum, plana	/ˈpleɪnəm/, /ˈpleɪnə/	A plateau or high plain e.g. Planum Boreum.	PM
Plume		A cryovolcanic feature on Triton. This term is currently unused.	PU
Promontorium, promontoria	/prɒmənˈtɔəriəm/, /prɒmənˈtɔəriə/	"Cape"; headland. Used only on the Moon.	PR
Regio, regiones	/ˈriːdʒioʊ/ ~ /ˈrɛdʒioʊ/, /rɛdʒiˈoʊniːz/	Large area marked by reflectivity or color distinctions from adjacent areas, or a broad geographic region	RE
Reticulum, reticula	/r□ˈtɪkjʊləm/, /r□ˈtɪkjʊlə/	reticular (netlike) pattern on Venus	RT
Rima, rimae	/ˈraɪmə/, /ˈraɪmiː/	Fissure. Used only on the Moon.	RI
Rupes, rupes	/ˈruːpiːz/	Scarp	RU
Satellite feature		A feature that shares the name of an associated feature, for example Hertzsprung D.	SF
Scopulus, scopuli	/ˈskɒpjʊlə/, /ˈskɒpjʊlaɪ/	Lobate or irregular scarp	SC
Serpens, serpentes	/ˈsɜːrpɛnz/, /sərˈpɛntiːz/	Sinuous feature with segments of positive and negative relief along its length	SE
Sinus	/ˈsaɪnəs/	"Bay"; small plain on Moon or Mars, e.g. Sinus Meridiani; On Titan, bay within bodies of liquid.	SI

Sulcus, sulci	/ˈsʌlkəs/, /ˈsʌlsaɪ/	Subparallel furrows and ridges	SU
Terra, terrae	/ˈtɛrə/, /ˈtɛriː/	Extensive land mass e.g. Arabia Terra, Aphrodite Terra.	TA
Tessera, tesserae	/ˈtɛsərə/, /ˈtɛsəriː/	An area of tile-like, polygonal terrain. This term is used only on Venus.	TE
Tholus, tholi	/ˈθoʊləs/, /ˈθoʊlaɪ/	Small domical mountain or hill e.g. Hecates Tholus.	TH
Undae	/ˈʌndiː/	A field of dunes. Used on Venus, Mars and Titan.	UN
Vallis, valles	/ˈvæl□s/, /ˈvæliːz/	A valley e.g. Valles Marineris.	VA
Vastitas, vastitates	/ˈvæst□təs/, /væst□ˈteɪtiːz/	An extensive plain. The only feature with this designation is Vastitas Borealis.	VS
Virga, virgae	/ˈvɜːrgə/, /ˈvɜːrdʒiː/	A streak or stripe of color. This term is currently used only on Titan.	VI

BY PLANET

- Geological features of the solar system
- Geological history of Earth
- Geology of Mercury
- Geology of Venus
- Geology of the Moon
- Geology of Mars
- Geology of Vesta
- Geology of Ceres
- Geology of Callisto
- Geology of Europa
- Geology of Ganymede
- Geology of Io
- Geology of Titan
- Geology of Triton
- Geology of Pluto
- Geology of Charon

GEOLOGY OF SOLAR TERRESTRIAL PLANETS

The **geology of solar terrestrial planets** mainly deals with the geological aspects of the four terrestrial planets of the Solar System, Mercury, Venus, Earth, and Mars, and one terrestrial dwarf planet, Ceres. Only one terrestrial planet, Earth, is known to have an active hydrosphere. Terrestrial planets are substantially different from the giant planets, which might not have solid surfaces and are composed mostly of some combination of hydrogen, helium, and water existing in various physical states. Terrestrial planets have a compact, rocky surfaces, and Venus, Earth, and Mars each also have an atmosphere. Their size, radius, and density are all similar.

The inner planets. From left to right: Mercury, Venus, Earth, Mars and terrestrial dwarf planet, Ceres (sizes to scale)

Terrestrial planets have numerous similarities to plutoids (objects like Pluto), which also have a solid surface, but are primarily composed of icy materials. During the formation of the Solar System, there were probably many more (planetesimals), but they have all merged with or been destroyed by the four remaining worlds in the solar nebula.

The Terrestrial planets all have roughly the same structure—a central metallic core, mostly iron, with a surrounding silicate mantle. The Moon is similar, but lacks a substantial iron core. Three of the four solar terrestrial planets (Venus, Earth, and Mars) have substantial atmospheres; all have impact craters and tectonic surface features such as rift valleys and volcanoes. The term *inner planet* should not be confused with *inferior planet*, which refers to those planets that are closer to the Sun than Earth is (i.e. Mercury and Venus).

Formation of Solar Planets

The Solar System is believed to have formed according to the nebular hypothesis, first proposed in 1755 by Immanuel Kant and independently formulated by Pierre-Simon Laplace. This theory holds that 4.6 billion years ago the Solar System formed from the gravitational collapse of a giant molecular cloud. This initial cloud was likely several light-years across and probably birthed several stars.

The first solid particles were microscopic in size. These particles orbited the Sun in nearly circular orbits right next to each other, as the gas from which they condensed. Gradually the gentle collisions allowed the flakes to stick together and make larger particles which, in turn, attracted more solid particles towards them. This process is known as accretion. The objects formed by accretion are called planetesimals—they act as seeds for planet formation. Initially, planetesimals were closely packed. They coalesced into larger objects, forming clumps of up to a few kilometers across in a few million years, a small time with comparison to the age of the Solar System. After the planetesimals grew bigger in sizes, collisions became highly destructive, making further growth more difficult. Only the biggest planetesimals survived the fragmentation process and continued to slowly grow into protoplanets by accretion of planetesimals of similar composition. After the protoplanet formed, accumulation of heat from radioactive decay of short-lived elements melted the planet, allowing materials to differentiate (i.e. to separate according to their density).

Terrestrial Planets

In the warmer inner Solar System, planetesimals formed from rocks and metals cooked billions of years ago in the cores of massive stars. These elements constituted only 0.6% of the material in the solar nebula. That is why the terrestrial planets could not grow very large and could not exert large pull on hydrogen and helium gas. Also, the faster collisions among particles close to the Sun were more destructive on average. Even if the terrestrial planets had had hydrogen and helium, the Sun would have heated the gases and caused them to escape. Hence, solar terrestrial planets such as Mercury, Venus, Earth, and Mars are dense small worlds composed mostly from 2% of heavier elements contained in the solar nebula.

Surface geology of inner solar planets

The four inner or terrestrial planets have dense, rocky compositions, few or no moons, and no ring systems. They are composed largely of minerals with high melting points, such as the silicates which form their solid crusts and semi-liquid mantles, and metals such as iron and nickel, which form their cores.

Mercury

The Mariner 10 mission (1974) mapped about half the surface of Mercury. On the basis of that data, scientists have a first-order understanding of the geology and history of the planet. Mercury's surface shows intercrater plains, basins, smooth plains, craters, and tectonic features.

Mercury›s oldest surface is its intercrater plains, which are present (but much less extensive) on the Moon. The intercrater plains are level to gently rolling terrain that occur between and around large craters. The plains predate the heavily cratered terrain, and have obliterated many of the early craters and basins of Mercury; they probably formed by widespread volcanism early in mercurian history.

Mercurian craters have the morphological elements of lunar craters—the smaller craters are bowl-shaped, and with increasing size, they develop scalloped rims, central peaks, and terraces on the inner walls. The ejecta sheets have a hilly, lineated texture and swarms of secondary impact craters. Fresh craters of all sizes have dark or bright halos and well-developed ray systems. Although mercurian and lunar craters are superficially similar, they show subtle differences, especially in deposit extent. The continuous ejecta and fields of secondary craters on Mercury are far less extensive (by a factor of about 0.65) for a given rim diameter than those of comparable lunar craters. This difference results from the 2.5 times higher gravitational field on Mercury compared with the Moon. As on the Moon, impact craters on Mercury are progressively degraded by subsequent impacts. The freshest craters have ray systems and a crisp morphology. With further degradation, the craters lose their crisp morphology and rays and features on the continuous ejecta become more blurred until only the raised rim near the crater remains recognizable. Because craters become progressively degraded with time, the degree of degradation gives a rough indication of the crater›s relative age. On the assumption that craters of similar size and morphology are roughly the same age, it is possible to place constraints on the ages of other underlying or overlying units and thus to globally map the relative age of craters.

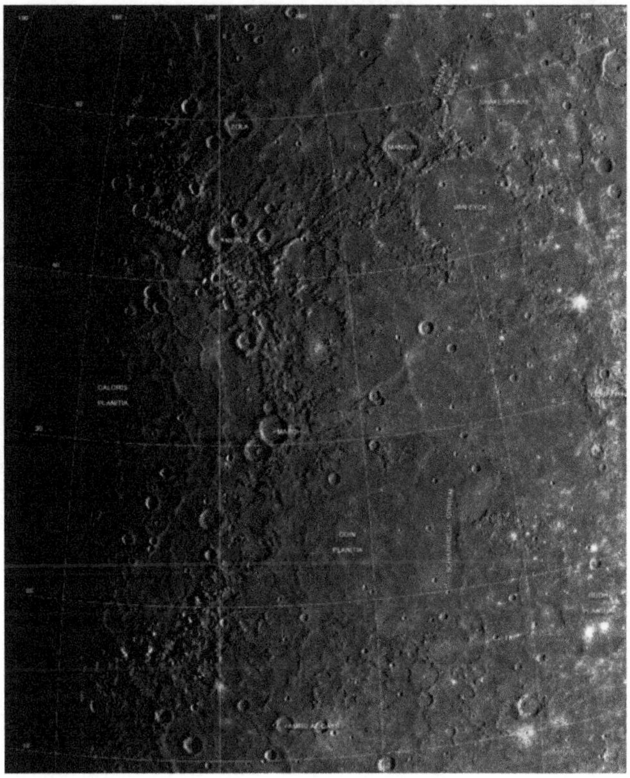

Mercury's Caloris Basin is one of the largest impact features in the Solar System.

At least 15 ancient basins have been identified on Mercury. Tolstoj is a true multi-ring basin, displaying at least two, and possibly as many as four, concentric rings. It has a well-preserved ejecta blanket extending outward as much as 500 kilometres (311 mi) from its rim. The basin interior is flooded with plains that clearly postdate the ejecta deposits. Beethoven has only one, subdued massif-like rim 625 kilometres (388 mi) in diameter, but displays an impressive, well lineated ejecta blanket that extends as far as 500 kilometres (311 mi). As at Tolstoj, Beethoven ejecta is asymmetric. The Caloris basin is defined by a ring of mountains 1,300 kilometres (808 mi) in diameter. Individual massifs are typically 30 kilometres (19 mi) to 50 kilometres (31 mi) long; the inner edge of the unit is marked by basin-facing scarps. Lineated terrain extends for about 1,000 kilometres (621 mi) out from the foot of a weak discontinuous scarp on the outer edge of the Caloris mountains; this terrain is similar to the *sculpture* surrounding the Imbrium basin on the Moon. Hummocky material forms a broad annulus about 800 kilometres (497 mi) from the Caloris mountains. It consists of low, closely spaced to scattered hills about 0.3 to 1 kilometre (1 mi) across and from tens of meters to a few hundred meters high. The outer boundary of this unit is gradational with the (younger) smooth plains that occur in the same region. A hilly and furrowed terrain is found antipodal to the Caloris basin, probably created by antipodal convergence of intense seismic waves generated by the Caloris impact.

The so-called "Weird Terrain" was formed by the Caloris Basin impact at its antipodal point.

The floor of the Caloris basin is deformed by sinuous ridges and fractures, giving the basin fill a grossly polygonal pattern. These plains may be volcanic, formed by the release of magma as part of the impact event, or a thick sheet of impact melt. Widespread areas of Mercury are covered by relatively flat, sparsely cratered plains materials. They fill depressions that range in size from regional troughs to crater floors. The smooth plains are similar to the maria of the Moon, an obvious difference being that the smooth plains have the same albedo as the intercrater plains. Smooth plains are most strikingly exposed in a broad annulus around the Caloris basin. No unequivocal volcanic features, such as flow lobes, leveed channels, domes, or cones are visible. Crater densities indicate that the smooth plains are significantly younger than ejecta from the Caloris basin. In addition, distinct color units, some of lobate shape, are observed in newly processed color data. Such relations strongly support a volcanic origin for the mercurian smooth plains, even in the absence of diagnostic landforms.

Lobate scarps are widely distributed over Mercury and consist of sinuous to arcuate scarps that transect preexisting plains and craters. They are most convincingly interpreted as thrust faults, indicating a period of global compression. The lobate scarps typically transect smooth plains materials (early Calorian age) on the floors of craters, but post-Caloris craters are superposed on them. These observations suggest that lobate-scarp formation was confined to a relatively narrow interval of time, beginning in the late pre-Tolstojan period and ending in the middle to late Calorian Period. In addition to scarps, wrinkle ridges occur in the smooth plains materials. These ridges probably were formed by local to regional surface compression caused by lithospheric loading by dense stacks of volcanic lavas, as suggested for those of the lunar maria.

Venus

The surface of Venus is comparatively very flat. When 93% of the topography was mapped by *Pioneer Venus*, scientists found that the total distance from the lowest point to the highest point on the entire surface was about 13 kilometres (8 mi), while on the Earth the distance from the basins to the Himalayas is about 20 kilometres (12.4 mi). According to the data of the altimeters of the *Pioneer*, nearly 51% of the surface is found located within 500 metres (1,640 ft) of the median radius of 6,052 km (3760 mi); only 2% of the surface is located at greater elevations than 2 kilometres (1 mi) from the median radius.

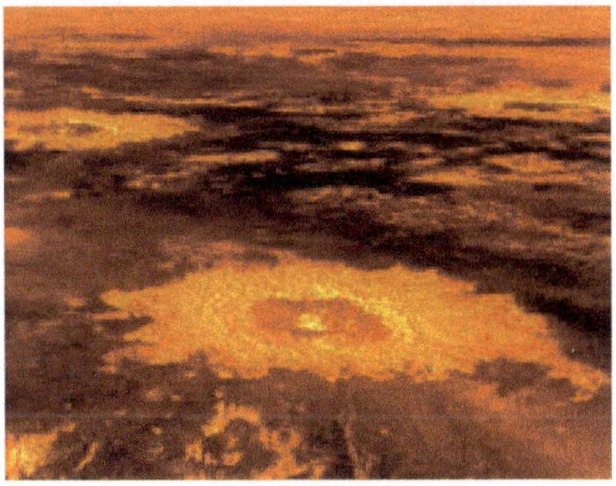

Danilova crater in relief

Venus shows no evidence of active plate tectonics. There is debatable evidence of active tectonics in the planet's distant past; however, events taking place since then (such as the plausible and generally accepted hypothesis that the Venusian lithosphere has thickened greatly over the course of several hundred million years) has made constraining the course of its geologic record difficult. However, the numerous well-preserved impact craters has been utilized as a dating method to approximately date the Venusian surface (since there are thus far no known samples of Venusian rock to be dated by more reliable methods). Dates derived are the dominantly in the range ~500 Mya–750Mya, although ages of up to ~1.2 Gya have been calculated. This research has led to the fairly well accepted hypothesis that Venus has undergone an essentially complete volcanic resurfacing at least once in its distant past, with the last event taking place approximately within the range of estimated surface ages. While the mechanism of such an impressionable thermal event remains a debated issue in Venusian geosciences, some scientists are advocates of processes involving plate motion to some extent. There are almost 1,000 impact craters on Venus, more or less evenly distributed across its surface.

Earth-based radar surveys made it possible to identify some topographic patterns related to craters, and the *Venera 15* and *Venera 16* probes identified almost 150 such features of probable impact origin. Global coverage from *Magellan* subsequently made it possible to identify nearly 900 impact craters.

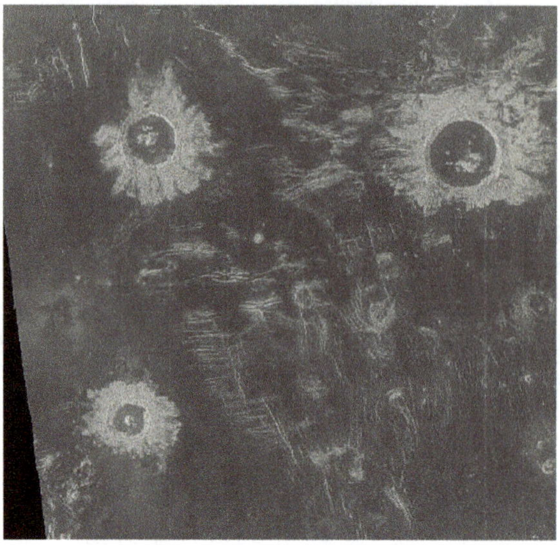

Danilova, Aglaonice and Saskja craters

Crater counts give an important estimate for the age of the surface of a planet. Over time, bodies in the Solar System are randomly impacted, so the more craters a surface has, the older it is. Compared to Mercury, the Moon and other such bodies, Venus has very few craters. In part, this is because Venus's dense atmosphere burns up smaller meteorites before they hit the surface. The *Venera* and *Magellan* data agree: there are very few impact craters with a diameter less than 30 kilometres (19 mi), and data from *Magellan* show an absence of any craters less than 2 kilometres (1 mi) in diameter. However, there are also fewer of the large craters, and those appear relatively young; they are rarely filled with lava, showing that they happened after volcanic activity in the area, and radar shows that they are rough and have not had time to be eroded down.

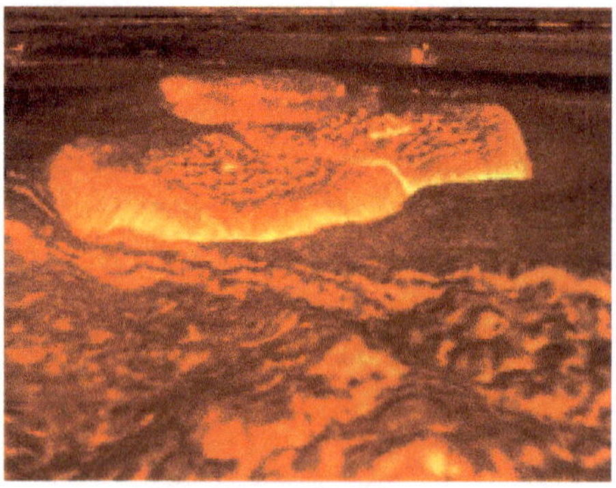

Computer generated perspective view of pancake domes in Venus's Alpha Regio

Much of Venus› surface appears to have been shaped by volcanic activity. Overall, Venus has several times as many volcanoes as Earth, and it possesses some 167 giant volcanoes that are over 100 kilometres (62 mi) across. The only volcanic complex of this size on Earth is the Big Island of Hawaii. However, this is not because Venus is more volcanically active than Earth, but because its crust is older. Earth›s crust is continually recycled by subduction at the boundaries of tectonic plates, and has an average age of about 100 million years, while Venus› surface is estimated to be about 500 million years old. Venusian craters range from 3 kilometres (2 mi) to 280 kilometres (174 mi) in diameter. There are no craters smaller than 3 km, because of the effects of the dense atmosphere on incoming objects. Objects with less than a certain kinetic energy are slowed down so much by the atmosphere that they do not create an impact crater.

Earth

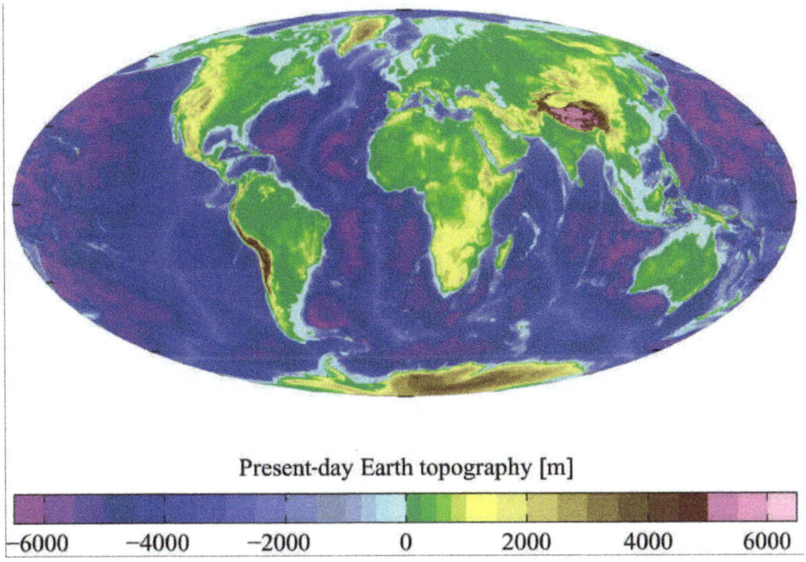

Present day Earth altimetry and bathymetry. Data from the National Geophysical Data Center›s TerrainBase Digital Terrain Model.

The Earth›s terrain varies greatly from place to place. About 70.8% of the surface is covered by water, with much of the continental shelf below sea level. The submerged surface has mountainous features, including a globe-spanning mid-ocean ridge system, as well as undersea volcanoes, oceanic trenches, submarine canyons, oceanic plateaus, and abyssal plains. The remaining 29.2% not covered by water consists of mountains, deserts, plains, plateaus, and other geomorphologies.

The planetary surface undergoes reshaping over geological time periods due to the effects of tectonics and erosion. The surface features built up or deformed through plate tectonics are subject to steady weathering from precipitation, thermal cycles, and chemical effects. Glaciation, coastal erosion, the build-up of coral reefs, and large meteorite impacts also act to reshape the landscape.

As the continental plates migrate across the planet, the ocean floor is subducted under the leading edges. At the same time, upwellings of mantle material create a divergent boundary along mid-ocean ridges. The combination of these processes continually recycles the ocean plate material. Most of the ocean floor is less than 100 million years in age.

The oldest ocean plate is located in the Western Pacific, and has an estimated age of about 200 million years. By comparison, the oldest fossils found on land have an age of about 3 billion years.

The continental plates consist of lower density material such as the igneous rocks granite and andesite. Less common is basalt, a denser volcanic rock that is the primary constituent of the ocean floors. Sedimentary rock is formed from the accumulation of sediment that becomes compacted together. Nearly 75% of the continental surfaces are covered by sedimentary rocks, although they form only about 5% of the crust. The third form of rock material found on Earth is metamorphic rock, which is created from the transformation of pre-existing rock types through high pressures, high temperatures, or both. The most abundant silicate minerals on the Earth›s surface include quartz, the feldspars, amphibole, mica, pyroxene, and olivine. Common carbonate minerals include calcite (found in limestone), aragonite, and dolomite.

The pedosphere is the outermost layer of the Earth that is composed of soil and subject to soil formation processes. It exists at the interface of the lithosphere, atmosphere, hydrosphere, and biosphere. Currently the total arable land is 13.31% of the land surface, with only 4.71% supporting permanent crops. Close to 40% of the Earth›s land surface is presently used for cropland and pasture, or an estimated 13 million square kilometres (5.0 million square miles) of cropland and 34 million square kilometres (13 million square miles) of pastureland.

The physical features of land are remarkably varied. The largest mountain ranges—the Himalayas in Asia and the Andes in South America—extend for thousands of kilometres. The longest rivers are the river Nile in Africa (6,695 kilometres or 4,160 miles) and the Amazon river in South America (6,437 kilometres or 4,000 miles). Deserts cover about 20% of the total land area. The largest is the Sahara, which covers nearly one-third of Africa.

The elevation of the land surface of the Earth varies from the low point of −418 m (−1,371 ft) at the Dead Sea, to a 2005-estimated maximum altitude of 8,848 m (29,028 ft) at the top of Mount Everest. The mean height of land above sea level is 686 m (2,250 ft).

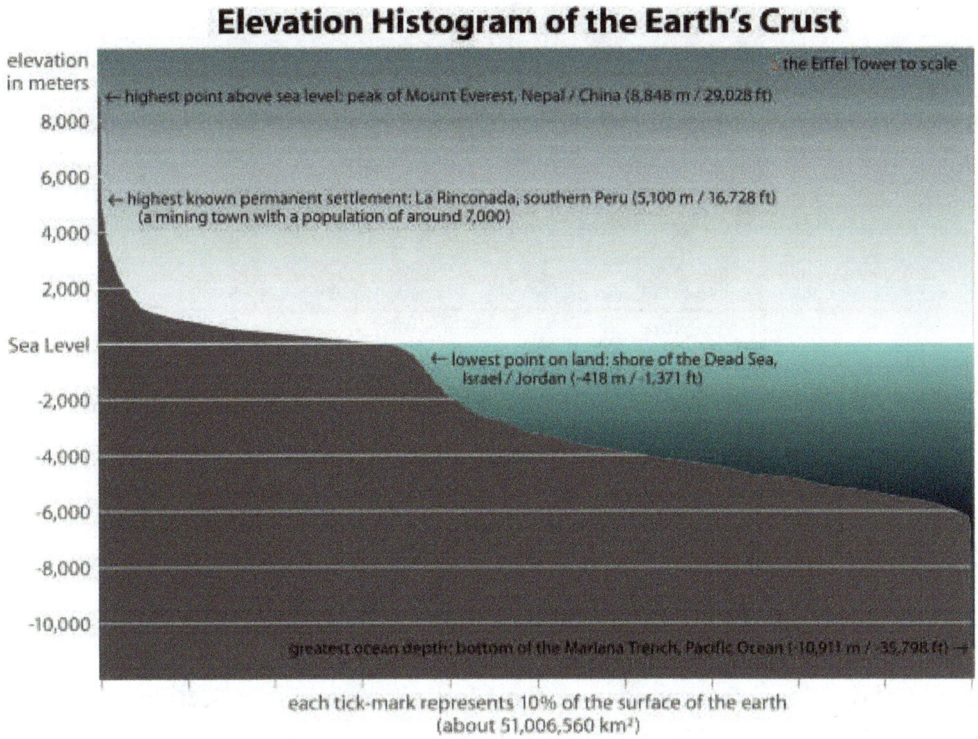

Elevation histogram of the surface of the Earth—approximately 71% of the Earth's surface is covered with water.

The geological history of Earth can be broadly classified into two periods namely:

- Precambrian: includes approximately 90% of geologic time. It extends from 4.6 billion years ago to the beginning of the Cambrian Period (about 570 Ma). It is generally believed that small proto-continents existed prior to 3000 Ma, and that most of the Earth›s landmasses collected into a single supercontinent around 1000 Ma.
- Phanerozoic: is the current eon in the geologic timescale. It covers roughly 545 million years. During the period covered, continents drifted about, eventually collected into a single landmass known as Pangea and then split up into the current continental landmasses.

Mars

Rock strewn surface imaged by Mars Pathfinder

The surface of Mars is thought to be primarily composed of basalt, based upon the observed lava flows from volcanos, the Martian meteorite collection, and data from landers and orbital observations. The lava flows from Martian volcanos show that that lava has a very low viscosity, typical of basalt. Analysis of the soil samples collected by the Viking landers in 1976 indicate iron-rich clays consistent with weathering of basaltic rocks. There is some evidence that some portion of the Martian surface might be more silica-rich than typical basalt, perhaps similar to andesitic rocks on Earth, though these observations may also be explained by silica glass, phyllosilicates, or opal. Much of the surface is deeply covered by dust as fine as talcum powder. The red/orange appearance of Mars' surface is caused by iron(III) oxide (rust). Mars has twice as much iron oxide in its outer layer as Earth does, despite their supposed similar origin. It is thought that Earth, being hotter, transported much of the iron downwards in the 1,800 kilometres (1,118 mi) deep, 3,200 °C (5,792 °F), lava seas of the early planet, while Mars, with a lower lava temperature of 2,200 °C (3,992 °F) was too cool for this to happen.

The core is surrounded by a silicate mantle that formed many of the tectonic and volcanic features on the planet. The average thickness of the planet's crust is about 50 km, and it is no thicker than 125 kilometres (78 mi), which is much thicker than Earth's crust which varies between 5 kilometres (3 mi) and 70 kilometres (43 mi). As a result, Mars' crust does not easily deform, as was shown by the recent radar map of the south polar ice cap which does not deform the crust despite being about 3 km thick.

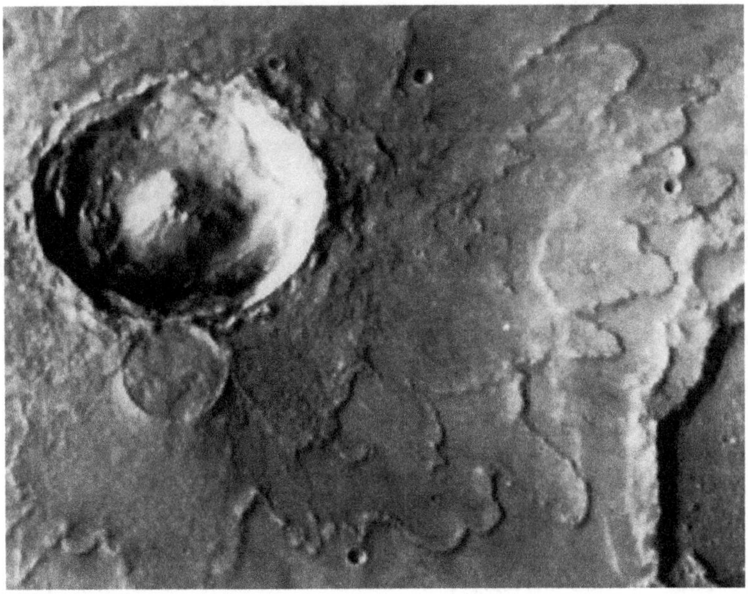

Yuty impact crater with typical rampart ejecta

Crater morphology provides information about the physical structure and composition of the surface. Impact craters allow us to look deep below the surface and into Mars geological past. Lobate ejecta blankets (pictured left) and central pit craters are common on Mars but uncommon on the Moon, which may indicate the presence of near-surface volatiles (ice and water) on Mars. Degraded impact structures record variations in volcanic, fluvial, and aeolian activity.

The Yuty crater is an example of a Rampart crater so called because of the rampart like edge of the ejecta. In the Yuty crater the ejecta completely covers an older crater at its side, showing that the ejected material is just a thin layer.

The geological history of Mars can be broadly classified into many epochs, but the following are the three major ones:

- Noachian epoch (named after Noachis Terra): Formation of the oldest extant surfaces of Mars, 3.8 billion years ago to 3.5 billion years ago. Noachian age surfaces are scarred by many large impact craters. The Tharsis bulge volcanic upland is thought to have formed during this period, with extensive flooding by liquid water late in the epoch.
- Hesperian epoch (named after Hesperia Planum): 3.5 billion years ago to 1.8 billion years ago. The Hesperian epoch is marked by the formation of extensive lava plains.
- Amazonian epoch (named after Amazonis Planitia): 1.8 billion years ago to present. Amazonian regions have few meteorite impact craters but are otherwise quite varied. Olympus Mons, the largest volcano in the known Universe, formed during this period along with lava flows elsewhere on Mars.

Ceres

The geology of the dwarf planet, Ceres, was largely unknown until Dawn spacecraft explored it in early 2015. However, certain surface features such as "Piazzi", named after the dwarf planets' discoverer, had been resolved.[a] Ceres's oblateness is consistent with a differentiated body, a rocky core overlain with an icy mantle. This 100-kilometer-thick mantle (23%–28% of Ceres by mass; 50% by volume) contains 200 million cubic kilometers of water, which is more than the amount of fresh water on Earth. This result is supported by the observations made by the Keck telescope in 2002 and by evolutionary modeling. Also, some characteristics of its surface and history (such as its distance from the Sun, which weakened solar radiation enough to allow some fairly low-freezing-point components to be incorporated during its formation), point to the presence of volatile materials in the interior of Ceres. It has been suggested that a remnant layer of liquid water may have survived to the present under a layer of ice. The surface composition of Ceres is broadly similar to that of C-type asteroids. Some differences do exist. The ubiquitous features of the Cererian IR spectra are those of hydrated materials, which indicate the presence of significant amounts of water in the interior. Other possible surface constituents include iron-rich clay minerals (cronstedtite) and carbonate minerals (dolomite and siderite), which are common minerals in carbonaceous chondrite meteorites. The spectral features of carbonates and clay minerals are usually absent in the spectra of other C-type asteroids. Sometimes Ceres is classified as a G-type asteroid.

The Cererian surface is relatively warm. The maximum temperature with the Sun overhead was estimated from measurements to be 235 K (about −38 °C, −36 °F) on 5 May 1991.

Prior to the Dawn mission, only a few Cererian surface features had been unambiguously detected. High-resolution ultraviolet Hubble Space Telescope images taken in 1995 showed a dark spot on its surface, which was nicknamed "Piazzi" in honor of the discoverer of Ceres. This was thought to be a crater. Later near-infrared images with a higher resolution taken over a whole rotation with the Keck telescope using adaptive optics showed several bright and dark features moving with Ceres's rotation. Two dark features had circular shapes and are presumably craters; one of them was observed to have a bright central region, whereas another was identified as the "Piazzi" feature. More recent visible-light Hubble Space Telescope images of a full rotation taken in 2003 and 2004 showed 11 recognizable surface features, the natures of which are currently unknown. One of these features corresponds to the "Piazzi" feature observed earlier.

These last observations also determined that the north pole of Ceres points in the direction of right ascension 19 h 24 min (291°), declination +59°, in the constellation Draco. This means that Ceres's axial tilt is very small—about 3°.

Atmosphere There are indications that Ceres may have a tenuous atmosphere and water frost on the surface. Surface water ice is unstable at distances less than 5 AU from the Sun, so it is expected to sublime if it is exposed directly to solar radiation. Water ice can migrate from the deep layers of Ceres to the surface, but escapes in a very short time. As a result, it is difficult to detect water vaporization. Water escaping from polar regions of Ceres was possibly observed in the early 1990s but this has not been unambiguously demonstrated. It may be possible to detect escaping water from the surroundings of a fresh impact crater or from cracks in the subsurface layers of Ceres. Ultraviolet observations by the IUE spacecraft detected statistically significant amounts of hydroxide ions near the Cererean north pole, which is a product of water-vapor dissociation by ultraviolet solar radiation.

In early 2014, using data from the Herschel Space Observatory, it was discovered that there are several localized (not more than 60 km in diameter) mid-latitude sources of water vapor on Ceres, which each give off about 10^{26} molecules (or 3 kg) of water per second. Two potential source regions, designated Piazzi (123°E, 21°N) and Region A (231°E, 23°N), have been visualized in the near infrared as dark areas (Region A also has a bright center) by the W. M. Keck Observatory. Possible mechanisms for the vapor release are sublimation from about 0.6 km2 of exposed surface ice, or cryovolcanic eruptions resulting from radiogenic internal heat or from pressurization of a subsurface ocean due to growth of an overlying layer of ice. Surface sublimation would be expected to decline as Ceres recedes from the Sun in its eccentric orbit, whereas internally powered emissions should not be affected by orbital position. The limited data available are more consistent with cometary-style sublimation. The spacecraft Dawn is approaching Ceres at aphelion, which may constrain Dawn's ability to observe this phenomenon.

Note: This info was taken directly from the main article, sources for the material are included there.

Small Solar System bodies

Asteroids, comets, and meteoroids are all debris remaining from the nebula in which the Solar System formed 4.6 billion years ago.

ASTEROID BELT

The asteroid belt is the circumstellar disc in the Solar System located roughly between the orbits of the planets Mars and Jupiter. It is occupied by numerous irregularly shaped bodies called asteroids or minor planets. The asteroid belt is also termed the main asteroid belt or main belt to distinguish it from other asteroid populations in the Solar System such as near-Earth asteroids and trojan asteroids. About half the mass of the belt is contained in the four largest asteroids: Ceres, Vesta, Pallas, and Hygiea. The total mass of the asteroid belt is approximately 4% that of the Moon, or 22% that of Pluto, and roughly twice that of Pluto›s moon Charon (whose diameter is 1200 km).

Ceres, the asteroid belt›s only dwarf planet, is about 950 km in diameter, whereas Vesta, Pallas, and Hygiea have mean diameters of less than 600 km. The remaining bodies range down to the size of a dust particle. The asteroid material is so thinly distributed that numerous unmanned spacecraft have traversed it without incident. Nonetheless, collisions between large asteroids do occur, and these can form an asteroid family whose members have similar orbital characteristics and compositions. Individual asteroids within the asteroid belt are categorized by their spectra, with most falling into three basic groups: carbonaceous (C-type), silicate (S-type), and metal-rich (M-type).

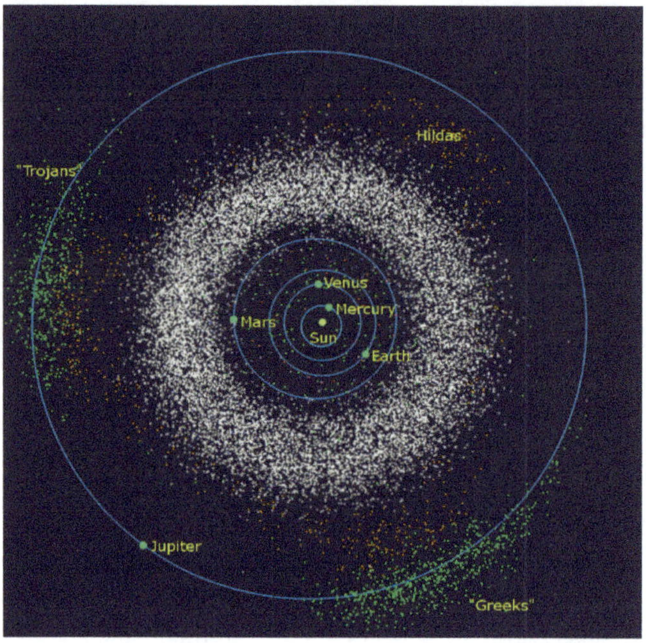

The asteroids of the inner Solar System and Jupiter: The donut-shaped asteroid belt is located between the orbits of Jupiter and Mars.

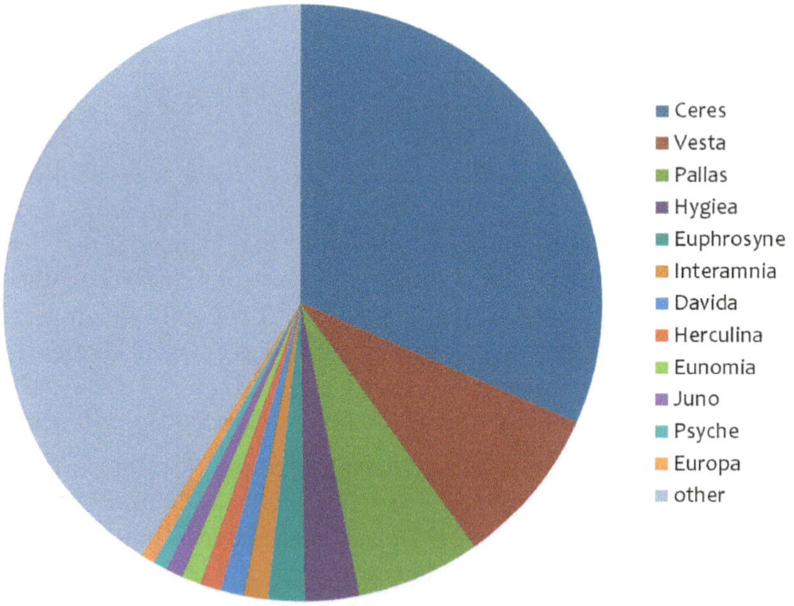

The relative masses of the top twelve asteroids known compared to the remaining mass of all the other asteroids in the belt.

By far the largest object within the belt is Ceres. The total mass of the asteroid belt is significantly less than Pluto's, and approximately twice that of Pluto's moon Charon.

The asteroid belt formed from the primordial solar nebula as a group of planetesimals. Planetesimals are the smaller precursors of the protoplanets. Between Mars and Jupiter, however, gravitational perturbations from Jupiter imbued the protoplanets with too much orbital energy for them to accrete into a planet. Collisions became too violent, and instead of fusing together, the planetesimals and most of the protoplanets shattered. As a result, 99.9% of the asteroid belt's original mass was lost in the first 100 million years of the Solar System's history. Some fragments eventually found their way into the inner Solar System, leading to meteorite impacts with the inner planets. Asteroid orbits continue to be appreciably perturbed whenever their period of revolution about the Sun forms an orbital resonance with Jupiter. At these orbital distances, a Kirkwood gap occurs as they are swept into other orbits.

Classes of small Solar System bodies in other regions are the near-Earth objects, the centaurs, the Kuiper belt objects, the scattered disc objects, the sednoids, and the Oort cloud objects.

On 22 January 2014, ESA scientists reported the detection, for the first definitive time, of water vapor on Ceres, the largest object in the asteroid belt. The detection was made by using the far-infrared abilities of the Herschel Space Observatory. The finding was unexpected because comets, not asteroids, are typically considered to sprout jets and plumes. According to one of the scientists, The lines are becoming more and more blurred between comets and asteroids.

In an anonymous footnote to his 1766 translation of Charles Bonnet's *Contemplation de la Nature*, the astronomer Johann Daniel Titius of Wittenberg noted an apparent pattern in the layout of the planets. If one began a numerical sequence at 0, then included 3, 6, 12, 24, 48, etc., doubling each time, and added four to each number and divided by 10, this produced a remarkably close approximation to the radii of the orbits of the known planets as measured in astronomical units. This pattern, now known as the Titius–Bode law, predicted the semi-major axes of the six planets of the time (Mercury, Venus, Earth, Mars, Jupiter and Saturn) provided one allowed for a "gap" between the orbits of Mars and Jupiter. In his footnote Titius declared, "But should the Lord Architect have left that space empty? Not at all." In 1768, the astronomer Johann Elert Bode made note of Titius's relationship in his *Anleitung zur Kenntniss des gestirnten Himmels* (English: *Instruction for the Knowledge of the Starry Heavens*) but did not credit Titius until later editions. It became known as "Bode's law". When William Herschel discovered Uranus in 1781, the planet's orbit matched the law almost perfectly, leading astronomers to conclude that there had to be a planet between the orbits of Mars and Jupiter.

Giuseppe Piazzi, discoverer of Ceres, the largest object in the asteroid belt. For several decades after its discovery Ceres was known as a planet, after which it was reclassified as asteroid number 1. In 2006 it was recognized to be a dwarf planet.

In 1800 the astronomer Baron Franz Xaver von Zach recruited 24 of his fellows into a club, the Vereinigte Astronomische Gesellschaft ("United Astronomical Society") which he informally dubbed the "Lilienthal Society"for its meetings in Lilienthal. Lilienthal is a town near Bremen, Germany that was home to a prominent telescope at the time. Determined to bring the Solar System to order, the group became known as the Himmelspolizei, or Celestial Police. Notable members included Herschel, the British astronomer royal, Nevil Maskelyne, Charles Messier, and Heinrich Olbers. The society assigned to each astronomer a 15° region of the zodiac to search for the missing planet.

Only a few months later, a non-member of the Celestial Police confirmed their expectations. On January 1, 1801, Giuseppe Piazzi, chair of astronomy at the University of Palermo, Sicily, found a tiny moving object in an orbit with exactly the radius predicted by the Titius–Bode law. He dubbed it Ceres, after the Roman goddess of the harvest and patron of Sicily. Piazzi initially believed it to be a comet, but its lack of a coma suggested it was a planet. Fifteen months later, Olbers discovered a second object in the same region, Pallas. Unlike the other known planets, the objects remained points of light even under the highest telescope magnifications instead of resolving into discs. Apart from their rapid movement, they appeared indistinguishable from stars. Accordingly, in 1802, William Herschel suggested they be placed into a separate category, named asteroids, after the Greek *asteroeides*, meaning "star-like". Upon completing a series of observations of Ceres and Pallas, he concluded,

Neither the appellation of planets, nor that of comets, can with any propriety of language be given to these two stars ... They resemble small stars so much as hardly to be distinguished from them. From this, their asteroidal appearance, if I take my name, and call them Asteroids; reserving for myself however the liberty of changing that name, if another, more expressive of their nature, should occur.

Despite Herschel's coinage, for several decades it remained common practice to refer to these objects as planets. By 1807, further investigation revealed two new objects in the region: 3 Juno and 4 Vesta. The burning of Lilienthal in the Napoleonic wars brought this first period of discovery to a close, and only in 1845 did astronomers detect another object (5 Astraea). Shortly thereafter, new objects were found at an accelerating rate, and counting them among the planets became increasingly cumbersome. Eventually, they were dropped from the planet list as first suggested by Alexander von Humboldt in the early 1850s, and Herschel's choice of nomenclature, "asteroids", gradually came into common use.

The discovery of Neptune in 1846 led to the discrediting of the Titius–Bode law in the eyes of scientists, because its orbit was nowhere near the predicted position. To date, there is no scientific explanation for the law, and astronomers' consensus regards it as a coincidence.

The expression asteroid belt came into use in the very early 1850s, although it is hard to pinpoint who coined the term. The first English use seems to be in the 1850 translation (by E. C. Otté) of Alexander von Humboldt›s *Cosmos*: "[...] and the regular appearance, about the 13th of November and the 11th of August, of shooting stars, which probably form part of a belt of asteroids intersecting the Earth's orbit and moving with planetary velocity". Other early appearances occur in Robert James Mann's *A Guide to the Knowledge of the Heavens*, "The orbits of the asteroids are placed in a wide belt of space, extending between the extremes of [...]". The American astronomer Benjamin Peirce seems to have adopted that terminology and to have been one of its promoters. One hundred asteroids had been located by mid-1868, and in 1891 the introduction of astrophotography by Max Wolf accelerated the rate of discovery still further. A total of 1,000 asteroids had been found by 1921, 10,000 by 1981, and 100,000 by 2000. Modern asteroid survey systems now use automated means to locate new minor planets in ever-increasing quantities.

Origin

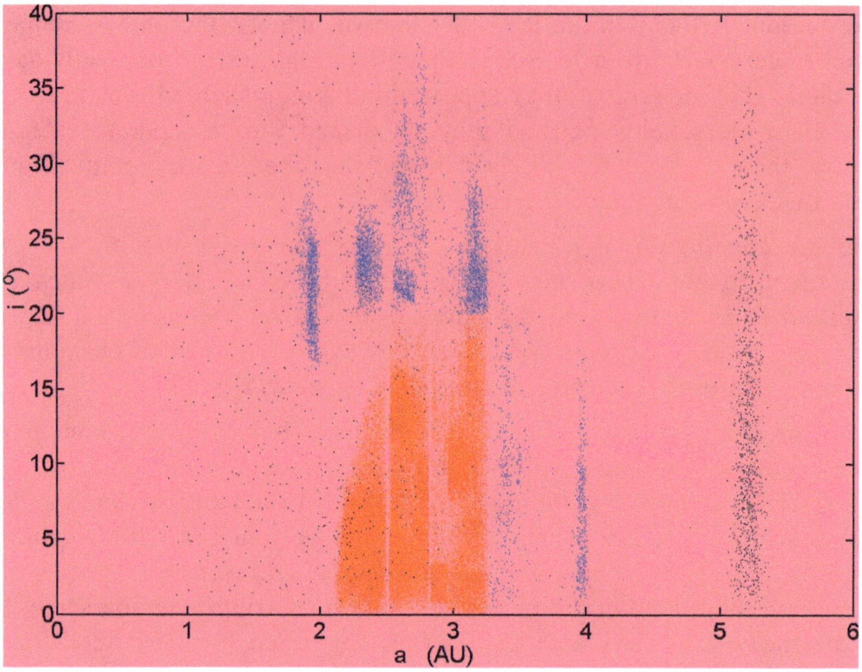

The asteroid belt showing the orbital inclinations versus distances from the Sun, with asteroids in the core region of the asteroid belt in red and other asteroids in blue

Formation

In 1802, shortly after discovering Pallas, Olbers suggested to Herschel that Ceres and Pallas were fragments of a much larger planet that once occupied the Mars–Jupiter region, this planet having suffered an internal explosion or a cometary impact many million years before. Over time, however, this hypothesis has fallen from favor. The large amount of energy required to destroy a planet, combined with the belt's low combined mass, which is only about 4% of the mass of the Moon, do not support the hypothesis. Further, the significant chemical differences between the asteroids become difficult to explain if they come from the same planet. Today, most scientists accept that, rather than fragmenting from a progenitor planet, the asteroids never formed a planet at all.

In general, in the Solar System, planetary formation is thought to have occurred via a process comparable to the long-standing nebular hypothesis: a cloud of interstellar dust and gas collapsed under the influence of gravity to form a rotating disc of material that then further condensed to form the Sun and planets. During the first few million years of the Solar System's history, an accretion process of sticky collisions caused the clumping of small particles, which gradually increased in size. Once the clumps reached sufficient mass, they could draw in other bodies through gravitational attraction and become planetesimals. This gravitational accretion led to the formation of the planets.

Planetesimals within the region which would become the asteroid belt were too strongly perturbed by Jupiter's gravity to form a planet. Instead they continued to orbit the Sun as before, occasionally colliding. In regions where the average velocity of the collisions was too high, the shattering of planetesimals tended to dominate over accretion, preventing the formation of planet-sized bodies. Orbital resonances occurred where the orbital period of an object in the belt formed an integer fraction of the orbital period of Jupiter, perturbing the object into a different orbit; the region lying between the orbits of Mars and Jupiter contains many such orbital resonances. As Jupiter migrated inward following its formation, these resonances would have swept across the asteroid belt, dynamically exciting the region's population and increasing their velocities relative to each other.

During the early history of the Solar System, the asteroids melted to some degree, allowing elements within them to be partially or completely differentiated by mass. Some of the progenitor bodies may even have undergone periods of explosive volcanism and formed magma oceans. However, because of the relatively small size of the bodies, the period of melting was necessarily brief (compared to the much larger planets), and had generally ended about 4.5 billion years ago, in the first tens of millions of years of formation. In August 2007, a study of zircon crystals in an Antarctic meteorite believed to have originated from 4 Vesta suggested that it, and by extension the rest of the asteroid belt, had formed rather quickly, within ten million years of the Solar System's origin.

Evolution

The asteroids are not samples of the primordial Solar System. They have undergone considerable evolution since their formation, including internal heating (in the first few tens of millions of years), surface melting from impacts, space weathering from radiation, and bombardment by micrometeorites. Although some scientists refer to the asteroids as residual planetesimals, other scientists consider them distinct.

The current asteroid belt is believed to contain only a small fraction of the mass of the primordial belt. Computer simulations suggest that the original asteroid belt may have contained mass equivalent to the Earth. Primarily because of gravitational perturbations, most of the material was ejected from the belt within about a million years of formation, leaving behind less than 0.1% of the original mass. Since their formation, the size distribution of the asteroid belt has remained relatively stable: there has been no significant increase or decrease in the typical dimensions of the main-belt asteroids.

The 4:1 orbital resonance with Jupiter, at a radius 2.06 AU, can be considered the inner boundary of the asteroid belt. Perturbations by Jupiter send bodies straying there into unstable orbits. Most bodies formed inside the radius of this gap were swept up by Mars (which has an aphelion at 1.67 AU) or ejected by its gravitational perturbations in the early history of the Solar System. The Hungaria asteroids lie closer to the Sun than the 4:1 resonance, but are protected from disruption by their high inclination.

When the asteroid belt was first formed, the temperatures at a distance of 2.7 AU from the Sun formed a snow line below the freezing point of water. Planetesimals formed beyond this radius were able to accumulate ice. In 2006 it was announced that a population of comets had been discovered within the asteroid belt beyond the snow line, which may have provided a source of water for Earth's oceans. According to some models, there was insufficient outgassing of water during the Earth's formative period to form the oceans, requiring an external source such as a cometary bombardment.

Characteristics

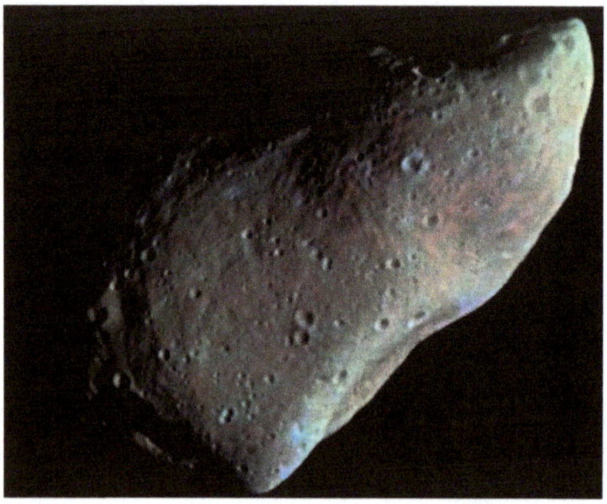

951 Gaspra, the first asteroid imaged by a spacecraft, as viewed during *Galileo's* 1991 flyby; colors are exaggerated

Fragment of the Allende meteorite, a carbonaceous chondrite that fell to Earth in Mexico in 1969

Contrary to popular imagery, the asteroid belt is mostly empty. The asteroids are spread over such a large volume that it would be improbable to reach an asteroid without aiming carefully. Nonetheless, hundreds of thousands of asteroids are currently known, and the total number ranges in the millions or more, depending on the lower size cutoff. Over 200 asteroids are known to be larger than 100 km, and a survey in the infrared wavelengths has shown that the asteroid belt has 0.7–1.7 million asteroids with a diameter of 1 km or more. The apparent magnitudes of most of the known asteroids are 11–19, with the median at about 16.

The total mass of the asteroid belt is estimated to be 2.8×10^{21} to 3.2×10^{21} kilograms, which is just 4% of the mass of the Moon. The four largest objects, Ceres, 4 Vesta, 2 Pallas, and 10 Hygiea, account for half of the belt›s total mass, with almost one-third accounted for by Ceres alone.

Composition

The current belt consists primarily of three categories of asteroids: C-type or carbonaceous asteroids, S-type or silicate asteroids, and M-type or metallic asteroids.

Carbonaceous asteroids, as their name suggests, are carbon-rich. They dominate the asteroid belt's outer regions. Together they comprise over 75% of the visible asteroids. They are more red in hue than the other asteroids and have a very low albedo. Their surface composition is similar to carbonaceous chondritemeteorites. Chemically, their spectra match the primordial composition of the early Solar System, with only the lighter elements and volatiles removed.

S-type (silicate-rich) asteroids are more common toward the inner region of the belt, within 2.5 AU of the Sun. The spectra of their surfaces reveal the presence of silicates and some metal, but no significant carbonaceous compounds. This indicates that their materials have been significantly modified from their primordial composition, probably through melting and reformation. They have a relatively high albedo, and form about 17% of the total asteroid population.

M-type (metal-rich) asteroids form about 10% of the total population; their spectra resemble that of iron-nickel. Some are believed to have formed from the metallic cores of differentiated progenitor bodies that were disrupted through collision. However, there are also some silicate compounds that can produce a similar appearance. For example, the large M-type asteroid 22 Kalliope does not appear to be primarily composed of metal. Within the asteroid belt, the number distribution of M-type asteroids peaks at a semi-major axis of about 2.7 AU. It is not yet clear whether all M-types are compositionally similar, or whether it is a label for several varieties which do not fit neatly into the main C and S classes.

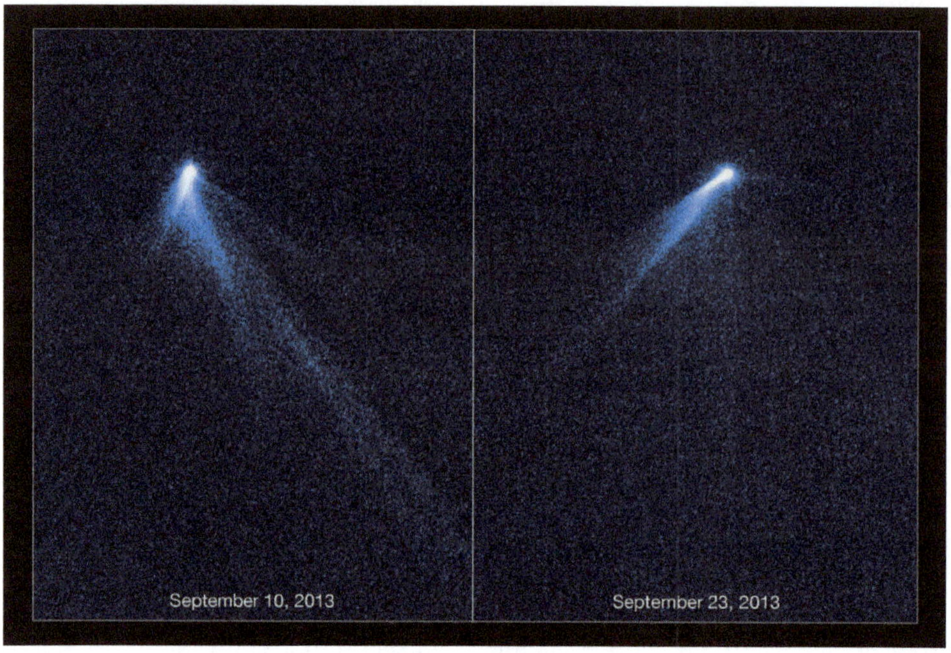

Hubble views extraordinary multi-tailed asteroid P/2013 P5.

One mystery of the asteroid belt is the relative rarity of V-type, or basaltic asteroids. Theories of asteroid formation predict that objects the size of Vesta or larger should form crusts and mantles, which would be composed mainly of basaltic rock, resulting in more than half of all asteroids being composed either of basalt or olivine. Observations, however, suggest that 99 percent of the predicted basaltic material is missing. Until 2001, most basaltic bodies discovered in the asteroid belt were believed to originate from the asteroid Vesta (hence their name V-type). However, the discovery of the asteroid 1459 Magnya revealed a slightly different chemical composition from the other basaltic asteroids discovered until then, suggesting a different origin. This hypothesis was reinforced by the further discovery in 2007 of two asteroids in the outer belt, 7472 Kumakiri and (10537) 1991 RY16, with differing basaltic composition that could not have originated from Vesta. These latter two are the only V-type asteroids discovered in the outer belt to date.

The temperature of the asteroid belt varies with the distance from the Sun. For dust particles within the belt, typical temperatures range from 200 K (−73 °C) at 2.2 AU down to 165 K (−108 °C) at 3.2 AU However, due to rotation, the surface temperature of an asteroid can vary considerably as the sides are alternately exposed to solar radiation and then to the stellar background.

Main-belt comets

Several otherwise unremarkable bodies in the outer belt show cometary activity. Because their orbits cannot be explained through capture of classical comets, it is thought that many of the outer asteroids may be icy, with the ice occasionally exposed to sublimation through small impacts. Main-belt comets may have been a major source of the Earth's oceans, because the deuterium–hydrogen ratio is too low for classical comets to have been the principal source.

Orbits

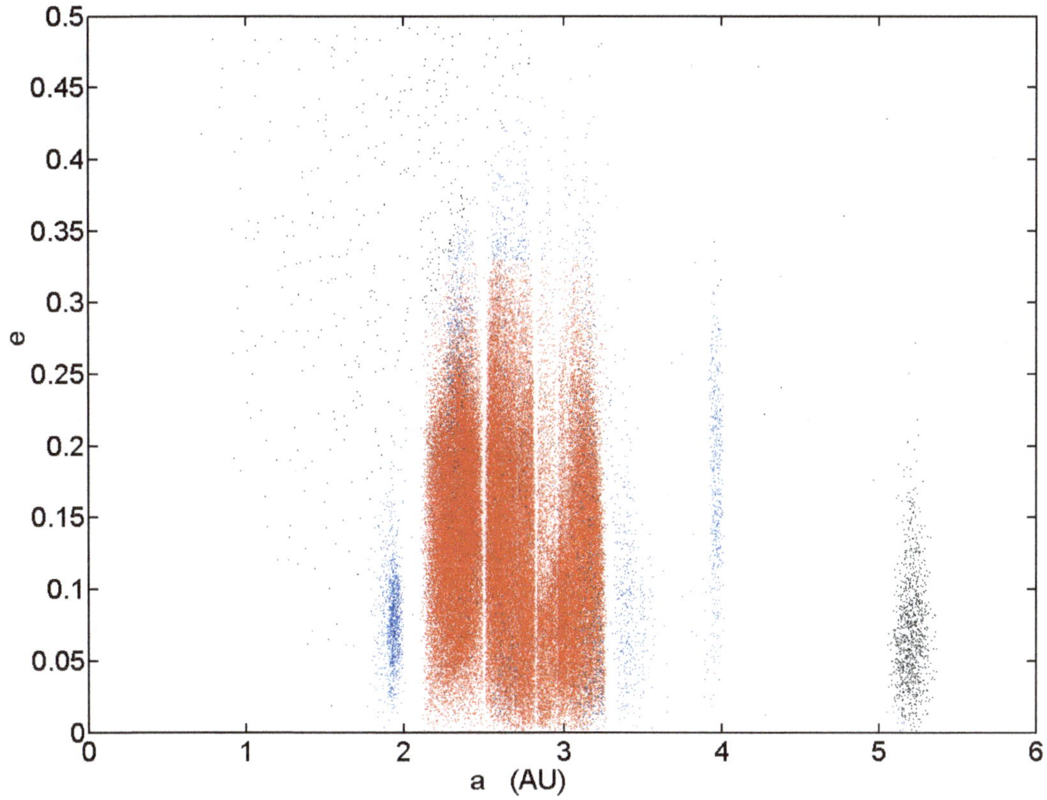

The asteroid belt (showing eccentricities), with the asteroid belt in red and blue ("core" region in red)

Most asteroids within the asteroid belt have orbital eccentricities of less than 0.4, and an inclination of less than 30°. The orbital distribution of the asteroids reaches a maximum at an eccentricity of around 0.07 and an inclination below 4°. Thus although a typical asteroid has a relatively circular orbit and lies near the plane of the ecliptic, some asteroid orbits can be highly eccentric or travel well outside the ecliptic plane.

Sometimes, the term *main belt* is used to refer only to the more compact "core" region where the greatest concentration of bodies is found. This lies between the strong 4:1 and 2:1 Kirkwood gaps at 2.06 and 3.27 AU, and at orbital eccentricities less than roughly 0.33, along with orbital inclinations below about 20°. This "core" region contains approximately 93.4% of all numbered minor planets within the Solar System.

Kirkwood Gaps

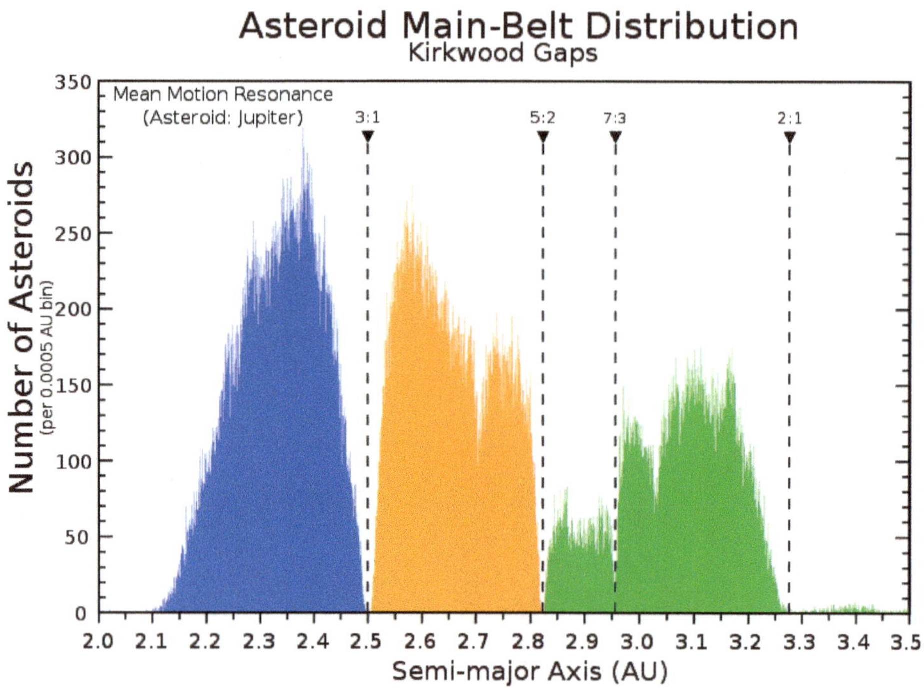

Number of asteroids in the asteroid belt as a function of their semi-major axis. The dashed lines indicate the Kirkwood gaps, where orbital resonances with Jupiter destabilize orbits. The color gives a possible division into three zones:

Zone I: inner main-belt (a < 2.5 AU)

Zone II: middle main-belt (2.5 AU < a > 2.82 AU)

Zone III: outer main-belt (a > 2.82 AU)

The semi-major axis of an asteroid is used to describe the dimensions of its orbit around the Sun, and its value determines the minor planet's orbital period. In 1866, Daniel Kirkwood announced the discovery of gaps in the distances of these bodies' orbits from the Sun. They were located at positions where their period of revolution about the Sun was an integer fraction of Jupiter›s orbital period. Kirkwood proposed that the gravitational perturbations of the planet led to the removal of asteroids from these orbits.

When the mean orbital period of an asteroid is an integer fraction of the orbital period of Jupiter, a mean-motion resonance with the gas giant is created that is sufficient to perturb an asteroid to new orbital elements. Asteroids that become located in the gap orbits (either primordially because of the migration of Jupiter›s orbit, or due to prior perturbations or collisions) are gradually nudged into different, random orbits with a larger or smaller semi-major axis.

The gaps are not seen in a simple snapshot of the locations of the asteroids at any one time because asteroid orbits are elliptical, and many asteroids still cross through the radii corresponding to the gaps. The actual spatial density of asteroids in these gaps does not differ significantly from the neighboring regions.

The main gaps occur at the 3:1, 5:2, 7:3, and 2:1 mean-motion resonances with Jupiter. An asteroid in the 3:1 Kirkwood gap would orbit the Sun three times for each Jovian orbit, for instance. Weaker resonances occur at other semi-major axis values, with fewer asteroids found than nearby. (For example, an 8:3 resonance for asteroids with a semi-major axis of 2.71 AU.)

The main or core population of the asteroid belt is sometimes divided into three zones, based on the most prominent Kirkwood gaps:

- Zone I lies between the 4:1 resonance (2.06 AU) and 3:1 resonance (2.5 AU) Kirkwood gaps.
- Zone II continues from the end of Zone I out to the 5:2 resonance gap (2.82 AU).
- Zone III extends from the outer edge of Zone II to the 2:1 resonance gap (3.28 AU).

The asteroid belt may also be divided into the inner and outer belts, with the inner belt formed by asteroids orbiting nearer to Mars than the 3:1 Kirkwood gap (2.5 AU), and the outer belt formed by those asteroids closer to Jupiter›s orbit. (Some authors subdivide the inner and outer belts at the 2:1 resonance gap (3.3 AU), whereas others suggest inner, middle, and outer belts; also see diagram).

Collisions

The zodiacal light, a minor part of which is created by dust from collisions in the asteroid belt

The high population of the asteroid belt makes for a very active environment, where collisions between asteroids occur frequently (on astronomical time scales). Collisions between main-belt bodies with a mean radius of 10 km are expected to occur about once every 10 million years. A collision may fragment an asteroid into numerous smaller pieces (leading to the formation of a new asteroid family). Conversely, collisions that occur at low relative speeds may also join two asteroids. After more than 4 billion years of such processes, the members of the asteroid belt now bear little resemblance to the original population. Along with the asteroid bodies, the asteroid belt also contains bands of dust with particle radii of up to a few hundred micrometres. This fine material is produced, at least in part, from collisions between asteroids, and by the impact of micrometeorites upon the asteroids. Due to the Poynting–Robertson effect, the pressure of solar radiation causes this dust to slowly spiral inward toward the Sun.

The combination of this fine asteroid dust, as well as ejected cometary material, produces the zodiacal light. This faint auroral glow can be viewed at night extending from the direction of the Sun along the plane of the ecliptic. Asteroid particles that produce the visible zodiacal light average about 40 μm in radius. The typical lifetimes of main-belt zodiacal cloud particles are about 700,000 years. Thus, to maintain the bands of dust, new particles must be steadily produced within the asteroid belt. It was once thought that collisions of asteroids form a major component of the zodiacal light. However, computer simulations by Nesvorný and colleagues attributed 85 percent of the zodiacal-light dust to fragmentations of Jupiter-family comets, rather than to comets and collisions between asteroids in the asteroid belt. At most 10 percent of the dust is attributed to the asteroid belt.

Meteorites

Some of the debris from collisions can form meteoroids that enter the Earth's atmosphere. Of the 50,000 meteorites found on Earth to date, 99.8 percent are believed to have originated in the asteroid belt.

Families and Groups

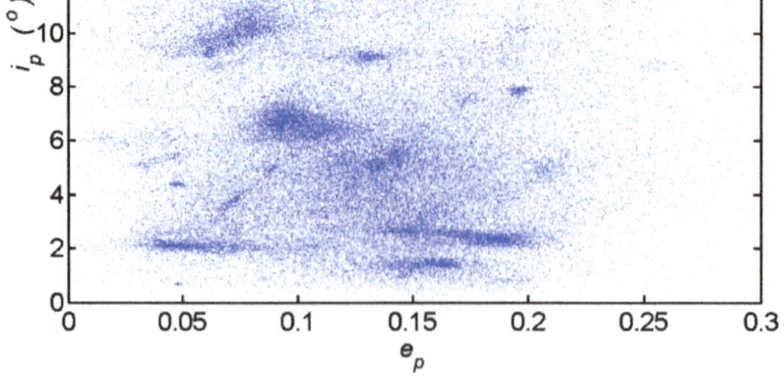

This plot of orbital inclination (i_p) versus eccentricity (e_p) for the numbered main-belt asteroids clearly shows clumpings representing asteroid families.

In 1918, the Japanese astronomer Kiyotsugu Hirayama noticed that the orbits of some of the asteroids had similar parameters, forming families or groups.

Approximately one-third of the asteroids in the asteroid belt are members of an asteroid family. These share similar orbital elements, such as semi-major axis, eccentricity, and orbital inclination as well as similar spectral features, all of which indicate a common origin in the breakup of a larger body. Graphical displays of these elements, for members of the asteroid belt, show concentrations indicating the presence of an asteroid family. There are about 20–30 associations that are almost certainly asteroid families. Additional groupings have been found that are less certain. Asteroid families can be confirmed when the members display common spectral features. Smaller associations of asteroids are called groups or clusters.

Some of the most prominent families in the asteroid belt (in order of increasing semi-major axes) are the Flora, Eunoma, Koronis, Eos, and Themis families. The Flora family, one of the largest with more than 800 known members, may have formed from a collision less than a billion years ago. The largest asteroid to be a true member of a family (as opposed to an interloper in the case of Ceres with the Gefion family) is 4 Vesta. The Vesta family is believed to have formed as the result of a crater-forming impact on Vesta. Likewise, the HED meteorites may also have originated from Vesta as a result of this collision.

Three prominent bands of dust have been found within the asteroid belt. These have similar orbital inclinations as the Eos, Koronis, and Themis asteroid families, and so are possibly associated with those groupings.

Periphery

Skirting the inner edge of the belt (ranging between 1.78 and 2.0 AU, with a mean semi-major axis of 1.9 AU) is the Hungaria family of minor planets. They are named after the main member, 434 Hungaria; the group contains at least 52 named asteroids. The Hungaria group is separated from the main body by the 4:1 Kirkwood gap and their orbits have a high inclination. Some members belong to the Mars-crossing category of asteroids, and gravitational perturbations by Mars are likely a factor in reducing the total population of this group.

Another high-inclination group in the inner part of the asteroid belt is the Phocaea family. These are composed primarily of S-type asteroids, whereas the neighboring Hungaria family includes some E-types. The Phocaea family orbit between 2.25 and 2.5 AU from the Sun.

Skirting the outer edge of the asteroid belt is the Cybele group, orbiting between 3.3 and 3.5 AU. These have a 7:4 orbital resonance with Jupiter. The Hilda family orbit between 3.5 and 4.2 AU, and have relatively circular orbits and a stable 3:2 orbital resonance with Jupiter. There are few asteroids beyond 4.2 AU, until Jupiter's orbit. Here the two families of Trojan asteroids can be found, which, at least for objects larger than 1 km, are approximately as numerous as the asteroids of the asteroid belt.

New families

Some asteroid families have formed recently, in astronomical terms. The Karin Cluster apparently formed about 5.7 million years ago from a collision with a progenitor asteroid 33 km in radius. The Veritas family formed about 8.3 million years ago; evidence includes interplanetary dust recovered from ocean sediment.

More recently, the Datura cluster appears to have formed about 450 thousand years ago from a collision with a main-belt asteroid. The age estimate is based on the probability of the members having their current orbits, rather than from any physical evidence. However, this cluster may have been a source for some zodiacal dust material. Other recent cluster formations, such as the Iannini cluster (*circa* 1–5 million years ago), may have provided additional sources of this asteroid dust.

Exploration

Artist's concept of the *Dawn* spacecraft with Vesta and Ceres

The first spacecraft to traverse the asteroid belt was *Pioneer 10*, which entered the region on 16 July 1972. At the time there was some concern that the debris in the belt would pose a hazard to the spacecraft, but it has since been safely traversed by 12 spacecraft without incident. *Pioneer 11*, *Voyagers 1* and *2* and *Ulysses* passed through the belt without imaging any asteroids. *Galileo* imaged 951 Gaspra in 1991 and 243 Ida in 1993, NEAR imaged 253 Mathilde in 1997, *Cassini* imaged 2685 Masursky in 2000, *Stardust* imaged 5535 Annefrank in 2002, *New Horizons* imaged 132524 APL in 2006, *Rosetta* imaged 2867 Šteins in September 2008 and 21 Lutetia in July 2010, and *Dawn* orbited Vesta between July 2011 and September 2012 and has orbited Ceres since March 2015. On its way to Jupiter, *Juno* traversed the asteroid belt without collecting science data. Due to the low density of materials within the belt, the odds of a probe running into an asteroid are now estimated at less than one in a billion.

Most belt asteroids imaged to date have come from brief flyby opportunities by probes headed for other targets. Only the *Dawn*, NEAR and *Hayabusa* missions have studied asteroids for a protracted period in orbit and at the surface. *Dawn* explored Vesta from July 2011 to September 2012, and has been orbiting Ceres since March 2015.

Comets

A comet is a small Solar System body that orbits the Sun and (at least occasionally) exhibits a coma (or atmosphere) and/or a tail—both primarily from the effects of solar radiation upon the comet's nucleus, which itself is a minor body composed of rock, dust, and ice.

Kuiper Belt

The Kuiper belt, sometimes called the Edgeworth–Kuiper belt, is a region of the Solar System beyond the planets extending from the orbit of Neptune (at 30 AU) to approximately 55 AU from the Sun. It is similar to the asteroid belt, although it is far larger; 20 times as wide and 20–200 times as massive. Like the asteroid belt, it consists mainly of small bodies (remnants from the Solar System's formation) and at least one dwarf planet—Pluto, wchich may be geologically active. But while the asteroid belt is composed primarily of rock and metal, the Kuiper belt is composed largely of ices, such as methane, ammonia, and water. The objects within the Kuiper belt, together with the members of the scattered disc and any potential Hills cloud or Oort cloud objects, are collectively referred to as trans-Neptunian objects (TNOs).

REFERENCES

1. Matt Williams (2015-08-23). What is the Asteroid Belt?. Universe Today. Retrieved 2016-01-30.
2. Krasinsky, G. A.; Pitjeva, E. V.; Vasilyev, M. V.; Yagudina, E. I. (July 2002). Hidden Mass in the Asteroid Belt. Icarus. 158 *(1): 98–105. Bibcode:2002Icar..158...98K. doi:10.1006/icar.2002.6837.*
3. Pitjeva, E. V. (2005). High-Precision Ephemerides of Planets—EPM and Determination of Some Astronomical Constants (PDF). Solar System Research. 39 *(3): 176–186. Bibcode:2005SoSyR..39..176P. doi:10.1007/s11208-005-0033-2. Archived from the original (PDF) on July 3, 2014.*
4. For recent estimates of the masses of Ceres, 4 Vesta, 2 Pallas and 10 Hygiea,

see the references in the infoboxes of their respective articles.

5. Yeomans, Donald K. (July 13, 2006). JPL Small-Body Database Browser. NASA JPL. Archived from the original on 29 September 2010. Retrieved 2010-09-27.
6. Brian Koberlein (2014-03-12). Why the Asteroid Belt Doesn›t Threaten Spacecraft. Universe Today. Retrieved 2016-01-30.
7. How Did The Asteroid Belt Form? Was There A Planet There?. CosmosUp. 2016-01-17. Retrieved 2016-01-30.
8. Nola Taylor Redd (2012-06-11). Asteroid Belt: Facts & Information. Space.com. Retrieved 2016-01-30.
9. Beatty, Kelly (March 10, 2009). Sculpting the Asteroid Belt. Sky & Telescope. Retrieved 2014-04-30.
10. Delgrande, J. J.; Soanes, S. V. (1943). Kirkwood›s Gap in the Asteroid Orbits. Journal of the Royal Astronomical Society of Canada. 37: 187. Bibcode:1943JRASC..37..187D.
11. Küppers, Michael; O'Rourke, Laurence; Bockelée-Morvan, Dominique; Zakharov, Vladimir; Lee, Seungwon; von Allmen, Paul; Carry, Benoît; Teyssier, David; Marston, Anthony; Müller, Thomas; Crovisier, Jacques; Barucci, M. Antonietta; Moreno, Raphael (2014). Localized sources of water vapour on the dwarf planet (1) Ceres. Nature. 505 *(7484): 525–527. Bibcode:2014Natur.505..525K. doi:10.1038/nature12918. ISSN 0028-0836. PMID 24451541.*
12. Harrington, J.D. (22 January 2014). Herschel Telescope Detects Water on Dwarf Planet – Release 14-021. NASA. Retrieved 22 January 2014.
13. Hilton, J. (2001). When Did the Asteroids Become Minor Planets?. US Naval Observatory (USNO). Retrieved 2007-10-01.
14. Dawn: A Journey to the Beginning of the Solar System. Space Physics Center: UCLA. 2005. Retrieved 2007-11-03.
15. Hoskin, Michael. Bode›s Law and the Discovery of Ceres. Churchill College, Cambridge. Retrieved 2010-07-12.
16. Linda T. Elkins-Tanton, Asteroids, Meteorites, and Comets, 2010:10
17. Call the police! The story behind the discovery of the asteroids. Astronomy Now (June 2007): 60–61.
18. Pogge, Richard (2006). An Introduction to Solar System Astronomy: Lecture 45: Is Pluto a Planet?. An Introduction to Solar System Astronomy. Ohio State University. Retrieved 2007-11-11.
19. Harper, Douglas (2010). Asteroid. Online Etymology Dictionary. Etymology Online. Retrieved 2011-04-15.
20. DeForest, Jessica (2000). Greek and Latin Roots. Michigan State University. Archived from the original on 12 August 2007. Retrieved 2007-07-25.
21. Cunningham, Clifford (1984). William Herschel and the First Two Asteroids. The Minor Planet Bulletin. Dance Hall Observatory, Ontario. 11: 3. Bibcode:1984MPBu...11....3C.

22. Staff (2002). Astronomical Serendipity. NASA JPL. Retrieved 2007-04-20.
23. Is it a coincidence that most of the planets fall within the Titius-Bode law›s boundaries?. astronomy.com. Retrieved 2014-01-22.
24. von Humboldt, Alexander (1850). Cosmos: A Sketch of a Physical Description of the Universe. 1. Harper & Brothers, New York (NY). p. 44. ISBN 0-8018-5503-9.
25. Mann, Robert James (1852). A Guide to the Knowledge of the Heavens. Jarrold. p. 171. and 1853, p. 216
26. Further Investigation relative to the form, the magnitude, the mass, and the orbit of the Asteroid Planets. The Edinburgh New Philosophical Journal. 5: 191. January–April 1857.: [Professor Peirce] then observed that the analogy between the ring of Saturn and the belt of the asteroids was worthy of notice.
27. Hughes, David W. (2007). A Brief History of Asteroid Spotting. BBC. Retrieved 2007-04-20.
28. Moore, Patrick; Rees, Robin (2011). Patrick Moore›s Data Book of Astronomy (2nd ed.). Cambridge University Press. p. 156. ISBN 0-521-89935-4.
29. Manley, Scott (August 25, 2010). Asteroid Discovery from 1980 to 2010. YouTube. Retrieved 2011-04-15.
30. MPC Archive Statistics. IAU Minor Planet Center. Retrieved 2011-04-04.
31. A Brief History of Asteroid Spotting. Open2.net. Retrieved 2007-05-15.
32. Masetti, M. & Mukai, K. (December 1, 2005). Origin of the Asteroid Belt. NASA Goddard Spaceflight Center. Retrieved 2007-04-25.
33. Watanabe, Susan (July 20, 2001). Mysteries of the Solar Nebula. NASA. Retrieved 2007-04-02.
34. Petit, J.-M.; Morbidelli, A. & Chambers, J. (2001). The Primordial Excitation and Clearing of the Asteroid Belt (PDF). Icarus. 153 *(2): 338–347. Bibcode:2001Icar..153..338P. doi:10.1006/icar.2001.6702. Archived (PDF) from the original on 21 February 2007. Retrieved 2007-03-22.*
35. Edgar, R. & Artymowicz, P. (2004). Pumping of a Planetesimal Disc by a Rapidly Migrating Planet (PDF). Monthly Notices of the Royal Astronomical Society. 354 *(3): 769–772. arXiv:astro-ph/0409017.* Bibcode:2004MNRAS.354..769E. doi:10.1111/j.1365-2966.2004.08238.x. Retrieved 2014-07-21.
36. Scott, E. r. d. (March 13–17, 2006). Constraints on Jupiter›s Age and Formation Mechanism and the Nebula Lifetime from Chondrites and Asteroids. Proceedings 37th Annual Lunar and Planetary Science Conference. League City, Texas: Lunar and Planetary Society. Retrieved 2007-04-16.
37. Taylor, G. J.; Keil, K.; McCoy, T.; Haack, H. & Scott, E. R. D. (1993). Asteroid differentiation – Pyroclastic volcanism to magma oceans. Meteoritics. 28 *(1): 34–52. Bibcode:1993Metic..28...34T. doi:10.1111/j.1945-5100.1993.tb00247.x.*
38. Kelly, Karen (2007). U of T researchers discover clues to early solar system. University of Toronto. Retrieved 2010-07-12.

39. Clark, B. E.; Hapke, B.; Pieters, C.; Britt, D. (2002). Asteroid Space Weathering and Regolith Evolution. Asteroids III. University of Arizona: 585. Bibcode:2002aste.conf..585C. Gaffey, Michael J. (1996). The Spectral and Physical Properties of Metal in Meteorite Assemblages: Implications for Asteroid Surface Materials. Icarus. 66 *(3): 468–486. Bibcode:1986Icar...66..468G. doi:10.1016/0019-1035(86)90086-2. ISSN 0019-1035.* Keil, K. (2000). Thermal alteration of asteroids: evidence from meteorites. Planetary and Space Science. Retrieved 2007-11-08. Baragiola, R. A.; Duke, C. A.; Loeffler, M.; McFadden, L. A. & Sheffield, J. (2003). Impact of ions and micrometeorites on mineral surfaces: Reflectance changes and production of atmospheric species in airless solar system bodies. EGS – AGU – EUG Joint Assembly: 7709. Bibcode:2003EAEJA.....7709B.

40. Chapman, C. R.; Williams, J. G.; Hartmann, W. K. (1978). The asteroids. Annual Review of Astronomy and Astrophysics. 16: 33–75. Bibcode:1978ARA&A..16...33C. doi:10.1146/annurev.aa.16.090178.000341.

41. Kracher, A. (2005). Asteroid 433 Eros and partially differentiated planetesimals: bulk depletion versus surface depletion of sulfur(PDF). Ames Laboratory. Archived (PDF) from the original on 28 November 2007. Retrieved 2007-11-08.

42. Robert Piccioni (2012-11-19). Did Asteroid Impacts Make Earth Habitable?. Guidetothecosmos.com. Retrieved 2013-05-03.

43. Stiles, Lori (September 15, 2005). Asteroids Caused the Early Inner Solar System Cataclysm. University of Arizona News. Retrieved 2007-04-18.

44. Alfvén, H.; Arrhenius, G. (1976). The Small Bodies. SP-345 Evolution of the Solar System. NASA. Archived from the original on 13 May 2007. Retrieved 2007-04-12.

45. Spratt, Christopher E. (April 1990). The Hungaria group of minor planets. Journal of the Royal Astronomical Society of Canada. 84: 123–131. Bibcode:1990JRASC..84..123S.

46. Lecar, M.; Podolak, M.; Sasselov, D.; Chiang, E. (2006). Infrared cirrus – New components of the extended infrared emission. The Astrophysical Journal. 640 *(2): 1115–1118. arXiv:astro-ph/0602217.* Bibcode:2006ApJ...640.1115L. doi:10.1086/500287.

47. Berardelli, Phil (March 23, 2006). Main-Belt Comets May Have Been Source Of Earths Water. Space Daily. Retrieved 2007-10-27.

48. Lakdawalla, Emily (April 28, 2006). Discovery of a Whole New Type of Comet. The Planetary Society. Archived from the original on 1 May 2007. Retrieved 2007-04-20.

49. Yeomans, Donald K. (April 26, 2007). JPL Small-Body Database Search Engine. NASA JPL. Retrieved 2007-04-26. – search for asteroids in the main belt regions with a diameter >100.

50. Tedesco, E. F. & Desert, F.-X. (2002). The Infrared Space Observatory Deep Asteroid Search. The Astronomical Journal. 123 *(4): 2070–2082. Bibcode:2002AJ....123.2070T. doi:10.1086/339482.*

51. Williams, Gareth (September 25, 2010). Distribution of the Minor Planets.

Minor Planets Center. Retrieved 2010-10-27.

52. Wiegert, P.; Balam, D.; Moss, A.; Veillet, C.; Connors, M. & Shelton, I. (2007). Evidence for a Color Dependence in the Size Distribution of Main-Belt Asteroids (PDF). The Astronomical Journal. 133 *(4): 1609–1614. arXiv:astro-ph/0611310.* Bibcode:2007AJ....133.1609W. doi:10.1086/512128. Retrieved 2008-09-06.

53. Clark, B. E. (1996). New News and the Competing Views of Asteroid Belt Geology. Lunar and Planetary Science. 27: 225–226. Bibcode:1996LPI....27..225C.

54. Margot, J. L. & Brown, M. E. (2003). A Low-Density M-type Asteroid in the Main Belt. Science. 300 *(5627): 1939–1942. Bibcode:2003Sci...300.1939M. doi:10.1126/science.1085844. PMID 12817147.*

55. Lang, Kenneth R. (2003). Asteroids and meteorites. NASA›s Cosmos. Retrieved 2007-04-02.

56. Mueller, M.; Harris, A. W.; Delbo, M. (2005). the MIRSI Team. 21 Lutetia and other M-types: Their sizes, albedos, and thermal properties from new IRTF measurements. Bulletin of the American Astronomical Society. 37: 627. Bibcode:2005DPS....37.0702M.

57. When is a comet not a comet?. ESA/Hubble Press Release. Retrieved 12 November 2013.

58. Duffard, R. D.; Roig, F. (July 14–18, 2008). Two New Basaltic Asteroids in the Main Belt?. Asteroids, Comets, Meteors 2008. Baltimore, Maryland. arXiv:0704.0230. Bibcode:2008LPICo1405.8154D.

59. Than, Ker (2007). Strange Asteroids Baffle Scientists. space.com. Retrieved 2007-10-14.

60. Low, F. J.; et al. (1984). Infrared cirrus – New components of the extended infrared emission. Astrophysical Journal Letters. 278: L19–L22. Bibcode:1984ApJ...278L..19L. doi:10.1086/184213.

61. Interview with David Jewitt. YouTube.com. 2007-01-05. Retrieved 2011-05-21.

62. This value was obtained by a simple count up of all bodies in that region using data for 120437 numbered minor planets from the Minor Planet Center orbit database, dated February 8, 2006.

63. Fernie, J. Donald (1999). The American Kepler. American Scientist. 87 *(5): 398. doi:10.1511/1999.5.398. Retrieved 2007-02-04.*

64. Liou, Jer-Chyi & Malhotra, Renu (1997). Depletion of the Outer Asteroid Belt. Science. 275 *(5298): 375–377. Bibcode:1997Sci...275..375L. doi:10.1126/science.275.5298.375. PMID 8994031. Retrieved 2007-08-01.*

65. McBride, N. & Hughes, D. W. (1990). The spatial density of asteroids and its variation with asteroidal mass. Monthly Notices of the Royal Astronomical Society. 244: 513–520. Bibcode:1990MNRAS.244..513M.

66. Ferraz-Mello, S. (June 14–18, 1993). Kirkwood Gaps and Resonant Groups. proceedings of the 160th International Astronomical Union. Belgirate, Italy:

Kluwer Academic Publishers. pp. 175–188. Retrieved 2007-03-28.

67. Klacka, Jozef (1992). Mass distribution in the asteroid belt. Earth, Moon, and Planets. 56 *(1): 47–52. Bibcode:1992EM&P...56...47K. doi:10.1007/BF00054599.*

68. Backman, D. E. (March 6, 1998). Fluctuations in the General Zodiacal Cloud Density. Backman Report. NASA Ames Research Center. Retrieved 2007-04-04.

69. David Nesvorný, William F. Bottke Jr, Luke Dones & Harold F. Levison (June 2002). The recent breakup of an asteroid in the main-belt region (PDF). Nature. 417: 720–722. Bibcode:2002Natur.417..720N. doi:10.1038/nature00789.

70. Reach, William T. (1992). Zodiacal emission. III – Dust near the asteroid belt. Astrophysical Journal. 392 *(1): 289–299. Bibcode:1992ApJ...392..289R. doi:10.1086/171428.*

71. Cometary origin of the zodiacal cloud and carbonaceous micrometeorites – Implications for hot debris disk

72. Kingsley, Danny (May 1, 2003). Mysterious meteorite dust mismatch solved. ABC Science. Retrieved 2007-04-04.

73. Meteors and Meteorites (PDF). NASA. Retrieved 2012-01-12.

74. Hughes, David W. (2007). Finding Asteroids In Space. BBC. Retrieved 2007-04-20.

75. Lemaitre, Anne (August 31 – September 4, 2004). Asteroid family classification from very large catalogues. Proceedings Dynamics of Populations of Planetary Systems. Belgrade, Serbia and Montenegro: Cambridge University Press. pp. 135–144. Retrieved 2007-04-15.

76. Martel, Linda M. V. (March 9, 2004). Tiny Traces of a Big Asteroid Breakup. Planetary Science Research Discoveries. Archived from the original on 1 April 2007. Retrieved 2007-04-02.

77. Drake, Michael J. (2001). The eucrite/Vesta story. Meteoritics & Planetary Science. 36 *(4): 501–513. Bibcode:2001M&PS...36..501D. doi:10.1111/j.1945-5100.2001.tb01892.x.*

78. Love, S. G. & Brownlee, D. E. (1992). The IRAS dust band contribution to the interplanetary dust complex – Evidence seen at 60 and 100 microns. Astronomical Journal. 104 *(6): 2236–2242. Bibcode:1992AJ....104.2236L. doi:10.1086/116399.*

79. Spratt, Christopher E. (1990). The Hungaria group of minor planets. Journal of the Royal Astronomical Society of Canada. 84 *(2): 123–131. Bibcode:1990JRASC..84..123S.*

80. Carvano, J. M.; Lazzaro, D.; Mothé-Diniz, T.; Angeli, C. A. & Florczak, M. (2001). Spectroscopic Survey of the Hungaria and Phocaea Dynamical Groups. Icarus. 149 *(1): 173–189. Bibcode:2001Icar..149..173C. doi:10.1006/icar.2000.6512.*

81. Dymock, Roger (2010). Asteroids and Dwarf Planets and How to Observe Them. Astronomers› Observing Guides. Springer. p. 24. ISBN 1-4419-6438-X. Retrieved 2011-04-04.

82. Nesvorný, David; et al. (August 2006). Karin cluster formation by asteroid impact. Icarus. 183 *(2): 296–311. Bibcode:2006Icar..183..296N. doi:10.1016/j.icarus.2006.03.008.*

83. McKee, Maggie (January 18, 2006). Eon of dust storms traced to asteroid smash. New Scientist Space. Retrieved 2007-04-15.

84. Nesvorný, D.; Vokrouhlick, D. & Bottke, W. F. (2006). The Breakup of a Main-Belt Asteroid 450 Thousand Years Ago. Science. 312 *(5779): 1490. Bibcode:2006Sci...312.1490N. doi:10.1126/science.1126175. PMID 16763141. Retrieved 2007-04-15.*

85. Nesvorný, D.; Bottke, W. F.; Levison, H. F. & Dones, L. (2003). Recent Origin of the Solar System Dust Bands (PDF). The Astrophysical Journal. 591 *(1): 486–497. Bibcode:2003ApJ...591..486N. doi:10.1086/374807. Retrieved 2007-04-15.*

86. Barucci, M. A.; Fulchignoni, M. & Rossi, A. (2007). Rosetta Asteroid Targets: 2867 Steins and 21 Lutetia. Space Science Reviews. 128 *(1–4): 67–78. Bibcode:2007SSRv..128...67B. doi:10.1007/s11214-006-9029-6.*

87. Greicius, Tony (July 31, 2015). NASA›s Juno Gives Starship-Like View of Earth Flyby. nasa.gov. NASA. Retrieved 4 September 2015.

88. Stern, Alan (June 2, 2006). New Horizons Crosses The Asteroid Belt. Space Daily. Retrieved 2007-04-14.

CHAPTER 8

Engineering Geology and Geotechnical Engineering

Chapter Contents

- **Engineering Geology**
- **History**
- **The Practice**
- **Scope of Studies**
- **Geohazards and Adverse Geological Conditions**
- **Soil and Rock Mechanics**
- **Methods and Reporting**
- **Geotechnical Engineering**

Engineering geologists are concerned with the detailed technical analysis of earth material and the risk assessment of geological hazards. Their role is to ensure that geological factors affecting engineering works are identified and dealt with.

They assess the integrity of soil, rock, groundwater and other natural conditions prior to major construction projects. They advise on procedures required for such developments and the suitability of appropriate construction materials.

Engineering geologists are also involved with analysing sites and designs for environmentally sensitive developments, such as landfill sites. By monitoring development areas and analysing ground conditions, they ensure that structures can be secure in the short and long term.

The term engineering geologist encompasses a range of roles and can be applied to many different sectors within the industry. It is only after working for a few years, and seeing how each department works, that engineering geologists are clear about which area they want to work in.

ENGINEERING GEOLOGY

Engineering geology is the application of the geological sciences to engineering study for the purpose of assuring that the geological factors regarding the location, design, construction, operation and maintenance of engineering works are recognized and accounted for. Engineering geologists provide geological and geotechnical recommendations, analysis, and design associated with human development and various types of structures. The realm of the engineering geologist is essentially in the area of earth-structure interactions, or investigation of how the earth or earth processes impact human made structures and human activities.

An engineering geologist logging rock core in the field, Western Australia.

Engineering geology studies may be performed during the planning, environmental impact analysis, civil or structural engineering design, value engineering and construction phases of public and private works projects, and during post-construction and forensic phases of projects. Works completed by engineering geologists include; geological hazard assessments, geotechnical, material properties, landslide and slope stability, erosion, flooding, dewatering, and seismic investigations, etc. Engineering geology studies are performed by a geologist or engineering geologist that is educated, trained and has obtained experience related to the recognition and interpretation of natural processes, the understanding of how these processes impact human made structures (and vice versa), and knowledge of methods by which to mitigate against hazards resulting from adverse natural or human made conditions. The principal objective of the engineering geologist is the protection of life and property against damage caused by various geological conditions.

The practice of engineering geology is also very closely related to the practice of geological engineering and geotechnical engineering. If there is a difference in the content of the disciplines, it mainly lies in the training or experience of the practitioner.

HISTORY

Although the study of geology has been around for centuries, at least in its modern form, the science and practice of engineering geology only commenced as a recognized discipline until the late 19th and early 20th centuries. The first book titled Engineering Geology was published in 1880 by William Penning. In the early 20th century Charles Berkey, an American trained geologist who was considered the first American engineering geologist, worked on several water-supply projects for New York City, then later worked on the Hoover dam and a multitude of other engineering projects. The first American engineering geology textbook was written in 1914 by Ries and Watson. In 1921 Reginald W. Brock, the first Dean of Applied Science at the University of British Columbia, started the first undergraduate and graduate degree programs in Geological Engineering, noting that students with an engineering foundation made first-class practicing geologists. In 1925, Karl Terzaghi, an Austrian trained engineer and geologist, published the first text in Soil Mechanics (in German). Terzaghi is known as the parent of soil mechanics, but also had great interest in geology; Terzaghi considered soil mechanics to be a sub-discipline of engineering geology. In 1929, Terzaghi, along with Redlich and Kampe, published their own Engineering Geology text (also in German).

The need for geologist on engineering works gained worldwide attention in 1928 with the failure of the St. Francis Dam in California and the death of 426 people. More engineering failures which occurred the following years also prompted the requirement for engineering geologists to work on large engineering projects.

In 1951, one of the earliest definitions of the "Engineering Geologist" or "Professional Engineering Geologist" was provided by the Executive Committee of the Division on Engineering Geology of the Geological Society of America.

THE PRACTICE

One of the most important roles as an engineering geologist is the interpretation of landforms and earth processes to identify potential geologic and related man-made hazards that may have a great impact on civil structures and human development.The background in geology provides the engineering geologist with an understanding of how the earth works, which is crucial minimizing earth related hazards. Most engineering geologists also have graduate degrees where they have gained specialized education and training in soil mechanics, rock mechanics, geotechnics, groundwater, hydrology, and civil design. These two aspects of the engineering geologists' education provides them with a unique ability to understand and mitigate for hazards associated with earth-structure interactions.

SCOPE OF STUDIES

Engineering geology investigation and studies may be performed:

- for residential, commercial and industrial developments;
- for governmental and military installations;
- for public works such as a stormwater drainage system, power plant, wind

turbine, transmission line, sewage treatment plant, water treatment plant, pipeline (aqueduct, sewer, outfall), tunnel, trenchless construction, canal, dam, reservoir, building foundation, railroad, transit, highway, bridge, seismic retrofit, power generation facility, airport and park;

- for mine and quarry developments, mine tailing dam, mine reclamation and mine tunneling;
- for wetland and habitat restoration programs;
- for government, commercial, or industrial hazardous waste remediation sites;
- for coastal engineering, sand replenishment, bluff or sea cliff stability, harbor, pier and waterfront development;
- for offshore outfall, drilling platform and sub-sea pipeline, sub-sea cable; and
- for other types of facilities.

GEOHAZARDS AND ADVERSE GEOLOGICAL CONDITIONS

Typical geologic hazards or other adverse conditions evaluated and mitigated by an engineering geologist include:

- fault rupture on seismically active faults ;
- seismic and earthquake hazards (ground shaking, liquefaction, lurching, lateral spreading, tsunami and seiche events);
- landslide, mudflow, rockfall, debris flow, and avalanche hazards ;
- unstable slopes and slope stability;
- erosion;
- slaking and heave of geologic formations, such as frost heaving;
- ground subsidence (such as due to ground water withdrawal, sinkhole collapse, cave collapse, decomposition of organic soils, and tectonic movement);
- volcanic hazards (volcanic eruptions, hot springs, pyroclastic flows, debris flow, debris avalanche, gas emissions, volcanic earthquakes);
- non-rippable or marginally rippable rock requiring heavy ripping or blasting;
- weak and collapsible soils, foundation bearing failures;
- shallow ground water/seepage; and
- other types of geologic constraints.

An engineering geologist or geophysicist may be called upon to evaluate the excavatability (i.e. rippability) of earth (rock) materials to assess the need for pre-blasting during earthwork construction, as well as associated impacts due to vibration during blasting on projects.

SOIL AND ROCK MECHANICS

Soil mechanics is a discipline that applies principles of engineering mechanics, e.g. kinematics, dynamics, fluid mechanics, and mechanics of material, to predict the mechanical behavior of soils. Rock mechanics is the theoretical and applied science of the mechanical behaviour of rock and rock masses; it is that branch of mechanics concerned with the response of rock and rock masses to the force-fields of their physical environment. The fundamental processes are all related to the behaviour of porous media. Together, soil and rock mechanics are the basis for solving many engineering geology problems.

METHODS AND REPORTING

The methods used by engineering geologists in their studies include

- geologic field mapping of geologic structures, geologic formations, soil units and hazards;
- the review of geologic literature, geologic maps, geotechnical reports, engineering plans, environmental reports, stereoscopic aerial photographs, remote sensing data, Global Positioning System (GPS) data, topographic maps and satellite imagery;
- the excavation, sampling and logging of earth/rock materials in drilled borings, backhoe test pits and trenches, fault trenching, and bulldozer pits;
- geophysical surveys (such as seismic refraction traverses, resistivity surveys, ground penetrating radar (GPR) surveys, magnetometer surveys, electromagnetic surveys, high-resolution sub-bottom profiling, and other geophysical methods);
- deformation monitoring as the systematic measurement and tracking of the alteration in the shape or dimensions of an object as a result of the application of stress to it manually or with an automatic deformation monitoring system; and
- other methods.

The fieldwork is typically culminated in analysis of the data and the preparation of an engineering geologic report, geotechnical report or design brief, fault hazard or seismic hazard report, geophysical report, ground water resource report or hydrogeologic report. The engineering geology report can also be prepared in conjunction with a geotechnical report, but commonly provides the same geotechnical analysis and design recommendations that would be presented in a geotechnical report. An engineering geology report describes the objectives, methodology, references cited, tests performed, findings and recommendations for development and detailed design of engineering works. Engineering geologists also provide geologic data on topographic maps, aerial photographs, geologic maps, Geographic Information System (GIS) maps, or other map bases.

GEOTECHNICAL ENGINEERING

Geotechnical engineering is the branch of civil engineering concerned with the engineering behavior of earth materials. Geotechnical engineering is important in civil engineering, but also has applications in military, mining, petroleum and other engineering disciplines that are concerned with construction occurring on the surface or within the ground. Geotechnical engineering uses principles of soil mechanics and rock mechanics to investigate subsurface conditions and materials; determine the relevant physical/mechanical and chemical properties of these materials; evaluate stability of natural slopes and man-made soil deposits; assess risks posed by site conditions; design earthworks and structure foundations; and monitor site conditions, earthwork and foundation construction.

Boston's Big Dig presented geotechnical challenges in an urban environment.

A typical geotechnical engineering project begins with a review of project needs to define the required material properties. Then follows a site investigation of soil, rock, fault distribution and bedrock properties on and below an area of interest to determine their engineering properties including how they will interact with, on or in a proposed construction. Site investigations are needed to gain an understanding of the area in or on which the engineering will take place. Investigations can include the assessment of the risk to humans, property and the environment from natural hazards such as earthquakes, landslides, sinkholes, soil liquefaction, debris flows and rockfalls.

A geotechnical engineer then determines and designs the type of foundations, earthworks, and/or pavement subgrades required for the intended man-made structures to be built. Foundations are designed and constructed for structures of various sizes such as high-rise buildings, bridges, medium to large commercial buildings, and smaller structures where the soil conditions do not allow code-based design.

Foundations built for above-ground structures include shallow and deep foundations. Retaining structures include earth-filled dams and retaining walls. Earthworks include embankments, tunnels, dikes and levees, channels, reservoirs, deposition of hazardous waste and sanitary landfills.

Geotechnical engineering is also related to coastal and ocean engineering. Coastal engineering can involve the design and construction of wharves, marinas, and jetties. Ocean engineering can involve foundation and anchor systems for offshore structures such as oil platforms.

The fields of geotechnical engineering and engineering geology are closely related, and have large areas of overlap. However, the field of geotechnical engineering is a specialty of engineering, where the field of engineering geology is a specialty of geology.

Humans have historically used soil as a material for flood control, irrigation purposes, burial sites, building foundations, and as construction material for buildings. First activities were linked to irrigation and flood control, as demonstrated by traces of dykes, dams, and canals dating back to at least 2000 BCE that were found in ancient Egypt, ancient Mesopotamia and the Fertile Crescent, as well as around the early settlements of Mohenjo Daro and Harappa in the Indus valley. As the cities expanded, structures were erected supported by formalized foundations; Ancient Greeks notably constructed pad footings and strip-and-raft foundations. Until the 18th century, however, no theoretical basis for soil design had been developed and the discipline was more of an art than a science, relying on past experience.

Several foundation-related engineering problems, such as the Leaning Tower of Pisa, prompted scientists to begin taking a more scientific-based approach to examining the subsurface. The earliest advances occurred in the development of earth pressure theories for the construction of retaining walls. Henri Gautier, a French Royal Engineer, recognized the «natural slope» of different soils in 1717, an idea later known as the soil›s angle of repose. A rudimentary soil classification system was also developed based on a material›s unit weight, which is no longer considered a good indication of soil type.

The application of the principles of mechanics to soils was documented as early as 1773 when Charles Coulomb (a physicist, engineer, and army Captain) developed improved methods to determine the earth pressures against military ramparts. Coulomb observed that, at failure, a distinct slip plane would form behind a sliding retaining wall and he suggested that the maximum shear stress on the slip plane, for design purposes, was the sum of the soil cohesion, c, and friction $\sigma \tan(\phi)$, where σ is the normal stress on the slip plane and ϕ is the friction angle of the soil. By combining Coulomb's theory with Christian Otto Mohr's 2D stress state, the theory became known as Mohr-Coulomb theory. Although it is now recognized that precise determination of cohesion is impossible because c is not a fundamental soil property, the Mohr-Coulomb theory is still used in practice today.

In the 19th century Henry Darcy developed what is now known as Darcy›s Law describing the flow of fluids in porous media. Joseph Boussinesq (a mathematician and physicist) developed theories of stress distribution in elastic solids that proved useful for estimating stresses at depth in the ground; William Rankine, an engineer and physicist, developed an alternative to Coulomb's earth pressure theory. Albert Atterberg developed the clay consistency indices that are still used today for soil classification. Osborne Reynolds recognized in 1885 that shearing causes volumetric dilation of dense and contraction of loose granular materials.

Modern geotechnical engineering is said to have begun in 1925 with the publication of *Erdbaumechanik* by Karl Terzaghi (a civil engineer and geologist). Considered by many to be the father of modern soil mechanics and geotechnical engineering, Terzaghi developed the principle of effective stress, and demonstrated that the shear strength of soil is controlled by effective stress. Terzaghi also developed the framework for theories of bearing capacity of foundations, and the theory for prediction of the rate of settlement of clay layers due to consolidation. In his 1948 book, Donald Taylor recognized that interlocking and dilation of densely packed particles contributed to the peak strength of a soil. The interrelationships between volume change behavior (dilation, contraction, and consolidation) and shearing behavior were all connected via the theory of plasticity using critical state soil mechanics by Roscoe, Schofield, and Wroth with the publication of «On the Yielding of Soils» in 1958. Critical state soil mechanics is the basis for many contemporary advanced constitutive models describing the behavior of soil.

Geotechnical centrifuge modeling is a method of testing physical scale models of geotechnical problems. The use of a centrifuge enhances the similarity of the scale model tests involving soil because the strength and stiffness of soil is very sensitive to the confining pressure. The centrifugal acceleration allows a researcher to obtain large (prototype-scale) stresses in small physical models.

Practicing Engineers

Geotechnical engineers are typically graduates of a four-year civil engineering program and some hold a masters degree. In the USA, geotechnical engineers are typically licensed and regulated as Professional Engineers (PEs) in most states; currently only California and Oregon have licensed geotechnical engineering specialties. The Academy of Geo-Professionals (AGP) began issuing Diplomate, Geotechnical Engineering (D.GE) certification in 2008. State governments will typically license engineers who have graduated from an ABET accredited school, passed the Fundamentals of Engineering examination, completed several years of work experience under the supervision of a licensed Professional Engineer, and passed the Professional Engineering examination.

Soil Mechanics

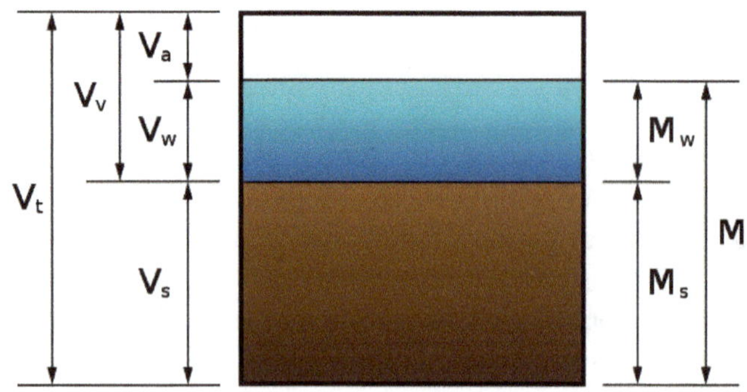

A phase diagram of soil indicating the weights and volumes of air, soil, water, and voids.

In geotechnical engineering, soils are considered a three-phase material composed of: rock or mineral particles, water and air. The voids of a soil, the spaces in between mineral particles, contain the water and air.

The engineering properties of soils are affected by four main factors: the predominant size of the mineral particles, the type of mineral particles, the grain size distribution, and the relative quantities of mineral, water and air present in the soil matrix. Fine particles (fines) are defined as particles less than 0.075 mm in diameter.

Soil Properties

Some of the important properties of soils that are used by geotechnical engineers to analyze site conditions and design earthworks, retaining structures, and foundations are:

Specific weight or Unit Weight

Cumulative weight of the solid particles, water and air of the unit volume of soil. Note that the air phase is often assumed to be weightless.

Porosity

Ratio of the volume of voids (containing air, water, or other fluids) in a soil to the total volume of the soil. Porosity is mathematically related to void ratio the by

$$n = \frac{e}{1+e}$$

Where e is void ratio and n is porosity

Void ratio

The ratio of the volume of voids to the volume of solid particles in a soil mass. Void ratio is mathematically related to the porosity by

$$e = \frac{n}{1-n}$$

Permeability

A measure of the ability of water to flow through the soil. It is expressed in units of velocity.

Compressibility

The rate of change of volume with effective stress. If the pores are filled with water, then the water must be squeezed out of the pores to allow volumetric compression of the soil; this process is called consolidation.

Shear strength

The maximum shear stress that can be applied in a soil mass without causing shear failure.

Atterberg Limits

Liquid limit, Plastic limit, and Shrinkage limit. These indices are used for estimation of other engineering properties and for soil classification.

Geotechnical Investigation

Geotechnical engineers and engineering geologists perform geotechnical investigations to obtain information on the physical properties of soil and rock underlying (and sometimes adjacent to) a site to design earthworks and foundations for proposed structures, and for repair of distress to earthworks and structures caused by subsurface conditions. A geotechnical investigation will include surface exploration and subsurface exploration of a site. Sometimes, geophysical methods are used to obtain data about sites. Subsurface exploration usually involves in-situ testing (two common examples of in-situ tests are the standard penetration test and cone penetration test). In addition site investigation will often include subsurface sampling and laboratory testing of the soil samples retrieved. The digging of test pits and trenching (particularly for locating faults and slide planes) may also be used to learn about soil conditions at depth. Large diameter borings are rarely used due to safety concerns and expense, but are sometimes used to allow a geologist or engineer to be lowered into the borehole for direct visual and manual examination of the soil and rock stratigraphy.

A variety of soil samplers exist to meet the needs of different engineering projects. The standard penetration test (SPT), which uses a thick-walled split spoon sampler, is the most common way to collect disturbed samples. Piston samplers, employing a thin-walled tube, are most commonly used for the collection of less disturbed samples. More advanced methods, such as ground freezing and the Sherbrooke block sampler, are superior, but even more expensive.

Atterberg limits tests, water content measurements, and grain size analysis, for example, may be performed on disturbed samples obtained from thick walled soil samplers. Properties such as shear strength, stiffness hydraulic conductivity, and coefficient of consolidation may be significantly altered by sample disturbance. To measure these properties in the laboratory, high quality sampling is required. Common tests to measure the strength and stiffness include the triaxial shear and unconfined compression test.

Surface exploration can include geologic mapping, geophysical methods, and photogrammetry; or it can be as simple as an engineer walking around to observe the physical conditions at the site. Geologic mapping and interpretation of geomorphology is typically completed in consultation with a geologist or engineering geologist.

Geophysical exploration is also sometimes used. Geophysical techniques used for subsurface exploration include measurement of seismic waves (pressure, shear, and Rayleigh waves), surface-wave methods and/or downhole methods, and electromagnetic surveys (magnetometer, resistivity, and ground-penetrating radar).

Foundations

A building's foundation transmits loads from buildings and other structures to the earth. Geotechnical engineers design foundations based on the load characteristics of the structure and the properties of the soils and/or bedrock at the site. In general, geotechnical engineers:

- Estimate the magnitude and location of the loads to be supported.
- Develop an investigation plan to explore the subsurface.
- Determine necessary soil parameters through field and lab testing (e.g., consolidation test, triaxial shear test, vane shear test, standard penetration test).
- Design the foundation in the safest and most economical manner.

The primary considerations for foundation support are bearing capacity, settlement, and ground movement beneath the foundations. Bearing capacity is the ability of the site soils to support the loads imposed by buildings or structures. Settlement occurs under all foundations in all soil conditions, though lightly loaded structures or rock sites may experience negligible settlements. For heavier structures or softer sites, both overall settlement relative to unbuilt areas or neighboring buildings, and differential settlement under a single structure, can be concerns. Of particular concern is settlement which occurs over time, as immediate settlement can usually be compensated for during construction. Ground movement beneath a structure's foundations can occur due to shrinkage or swell of expansive soils due to climatic changes, frost expansion of soil, melting of permafrost, slope instability, or other causes. All these factors must be considered during design of foundations. Many building codes specify basic foundation design parameters for simple conditions, frequently varying by jurisdiction, but such design techniques are normally limited to certain types of construction and certain types of sites, and are frequently very conservative. In areas of shallow bedrock, most foundations may bear directly on bedrock; in other areas, the soil may provide sufficient strength for the support of structures. In areas of deeper bedrock with soft overlying soils, deep foundations are used to support structures directly on the bedrock; in areas where bedrock is not economically available, stiff "bearing layers" are used to support deep foundations instead.

Shallow Foundations

Example of a slab-on-grade foundation.

Shallow foundations are a type of foundation that transfers building load to the very near the surface, rather than to a subsurface layer. Shallow foundations typically have a depth to width ratio of less than 1.

Footings

Footings (often called "spread footings" because they spread the load) are structural elements which transfer structure loads to the ground by direct areal contact. Footings can be isolated footings for point or column loads, or strip footings for wall or other long (line) loads. Footings are normally constructed from reinforced concrete cast directly onto the soil, and are typically embedded into the ground to penetrate through the zone of frost movement and/or to obtain additional bearing capacity.

Slab foundations

A variant on spread footings is to have the entire structure bear on a single slab of concrete underlying the entire area of the structure. Slabs must be thick enough to provide sufficient rigidity to spread the bearing loads somewhat uniformly, and to minimize differential settlement across the foundation. In some cases, flexure is allowed and the building is constructed to tolerate small movements of the foundation instead. For small structures, like single-family houses, the slab may be less than 300 mm thick; for larger structures, the foundation slab may be several meters thick.

Slab foundations can be either slab-on-grade foundations or embedded foundations, typically in buildings with basements. Slab-on-grade foundations must be designed to allow for potential ground movement due to changing soil conditions.

Deep foundations

Deep foundations are used for structures or heavy loads when shallow foundations cannot provide adequate capacity, due to size and structural limitations. They may also be used to transfer building loads past weak or compressible soil layers. While shallow foundations rely solely on the bearing capacity of the soil beneath them, deep foundations can rely on end bearing resistance, frictional resistance along their length, or both in developing the required capacity. Geotechnical engineers use specialized tools, such as the cone penetration test, to estimate the amount of skin and end bearing resistance available in the subsurface.

There are many types of deep foundations including piles, drilled shafts, caissons, piers, and earth stabilized columns. Large buildings such as skyscrapers typically require deep foundations. For example, the Jin Mao Tower in China uses tubular steel piles about 1m (3.3 feet) driven to a depth of 83.5m (274 feet) to support its weight.

Pile-driving for a bridge in Napa, California.

In buildings that are constructed and found to undergo settlement, underpinning piles can be used to stabilise the existing building.

There are three ways to place piles for a deep foundation. They can be driven, drilled, or installed by use of an auger. Driven piles are extended to their necessary depths with the application of external energy in the same way a nail is hammered. There are four typical hammers used to drive such piles: drop hammers, diesel hammers, hydraulic hammers, and air hammers. Drop hammers simply drop a heavy weight onto the pile to drive it, while diesel hammers use a single cylinder diesel engine to force piles through the Earth. Similarly, hydraulic and air hammers supply energy to piles through hydraulic and air forces. Energy imparted from a hammer head varies with type of hammer chosen, and can be as high as a million foot pounds for large scale diesel hammers, a very common hammer head used in practice. Piles are made of a variety of material including steel, timber, and concrete. Drilled piles are created by first drilling a hole to the appropriate depth, and filling it with concrete. Drilled piles can typically carry more load than driven piles, simply due to a larger diameter pile. The auger method of pile installation is similar to drilled pile installation, but concrete is pumped into the hole as the auger is being removed.

Lateral earth Support Structures

A retaining wall is a structure that holds back earth. Retaining walls stabilize soil and rock from downslope movement or erosion and provide support for vertical or near-vertical grade changes. Cofferdams and bulkheads, structures to hold back water, are sometimes also considered retaining walls.

The primary geotechnical concern in design and installation of retaining walls is that the weight of the retained material is creates lateral earth pressure behind the wall, which can cause the wall to deform or fail. The lateral earth pressure depends on the height of the wall, the density of the soil,the strength of the soil, and the amount of allowable movement of the wall. This pressure is smallest at the top and increases toward the bottom in a manner similar to hydraulic pressure, and tends to push the wall away from the backfill. Groundwater behind the wall that is not dissipated by a drainage system causes an additional horizontal hydraulic pressure on the wall.

Gravity Walls

Gravity walls depend on the size and weight of the wall mass to resist pressures from behind. Gravity walls will often have a slight setback, or batter, to improve wall stability. For short, landscaping walls, gravity walls made from dry-stacked (mortarless) stone or segmental concrete units (masonry units) are commonly used.

Earlier in the 20th century, taller retaining walls were often gravity walls made from large masses of concrete or stone. Today, taller retaining walls are increasingly built as composite gravity walls such as: geosynthetic or steel-reinforced backfill soil with precast facing; gabions (stacked steel wire baskets filled with rocks), crib walls (cells built up log cabin style from precast concrete or timber and filled with soil or free draining gravel) or soil-nailed walls (soil reinforced in place with steel and concrete rods).

For **reinforced-soil gravity walls**, the soil reinforcement is placed in horizontal layers throughout the height of the wall. Commonly, the soil reinforcement is **geogrid**, a high-strength polymer mesh, that provide tensile strength to hold soil together. The wall face is often of precast, segmental concrete units that can tolerate some differential movement. The reinforced soil's mass, along with the facing, becomes the gravity wall. The reinforced mass must be built large enough to retain the pressures from the soil behind it. Gravity walls usually must be a minimum of 30 to 40 percent as deep (thick) as the height of the wall, and may have to be larger if there is a slope or surcharge on the wall.

Cantilever Walls

Prior to the introduction of modern reinforced-soil gravity walls, cantilevered walls were the most common type of taller retaining wall. Cantilevered walls are made from a relatively thin stem of steel-reinforced, cast-in-place concrete or mortared masonry (often in the shape of an inverted T). These walls cantilever loads (like a beam) to a large, structural footing; converting horizontal pressures from behind the wall to vertical pressures on the ground below. Sometimes cantilevered walls are buttressed on the front, or include a counterfort on the back, to improve their stability against high loads. Buttresses are short wing walls at right angles to the main trend of the wall. These walls require rigid concrete footings below seasonal frost depth. This type of wall uses much less material than a traditional gravity wall.

Cantilever walls resist lateral pressures by friction at the base of the wall and/or **passive earth pressure**, the tendency of the soil to resist lateral movement.

Basements are a form of cantilever walls, but the forces on the basement walls are greater than on conventional walls because the basement wall is not free to move.

Excavation Shoring

Shoring of temporary excavations frequently requires a wall design which does not extend laterally beyond the wall, so shoring extends below the planned base of the excavation. Common methods of shoring are the use of **sheet piles** or **soldier beams and lagging**. Sheet piles are a form of driven piling using thin interlocking sheets of steel to obtain a continuous barrier in the ground, and are driven prior to excavation. Soldier beams are constructed of wide flange steel H sections spaced about 2–3 m apart, driven prior to excavation. As the excavation proceeds, horizontal timber or steel sheeting (lagging) is inserted behind the H pile flanges.

In some cases, the lateral support which can be provided by the shoring wall alone is insufficient to resist the planned lateral loads; in this case additional support is provided by walers or tie-backs. Walers are structural elements which connect across the excavation so that the loads from the soil on either side of the excavation are used to resist each other, or which transfer horizontal loads from the shoring wall to the base of the excavation. Tie-backs are steel tendons drilled into the face of the wall which extend beyond the soil which is applying pressure to the wall, to provide additional lateral resistance to the wall.

Earthworks

Excavation

Excavation is the process of training earth according to requirement by removing the soil from the site.

Filling

Filling is the process of training earth according to requirement by placing the soil on the site.

Compaction

Compaction is the process by which the density of soil is increased and permeability of soil is decreased. Fill placement work often has specifications requiring a specific degree of compaction, or alternatively, specific properties of the compacted soil. In-situ soils can be compacted by rolling, deep dynamic compaction, vibration, blasting, gyrating, kneading, compaction grouting etc.

A compactor/roller operated by U.S. Navy Seabees

Ground Improvement

Ground Improvement is a technique that improves the engineering properties of the treated soil mass. Usually, the properties modified are shear strength, stiffness and permeability. Ground improvement has developed into a sophisticated tool to support foundations for a wide variety of structures. Properly applied, i.e. after giving due consideration to the nature of the ground being improved and the type and sensitivity of the structures being built, ground improvement often reduces direct costs and saves time.

Slope stabilization

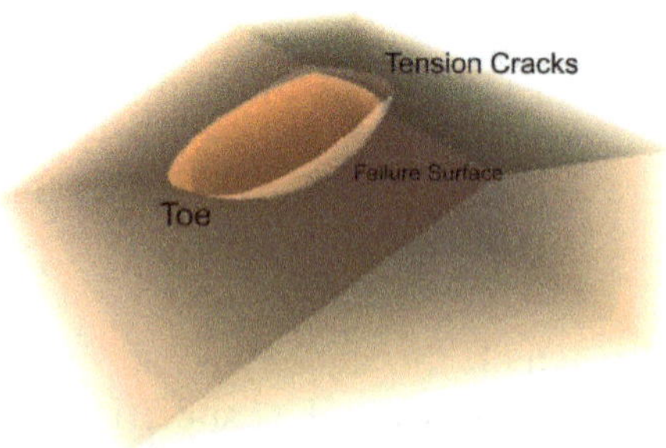

Simple slope slip section.

Slope stability is the potential of soil covered slopes to withstand and undergo movement. Stability is determined by the balance of shear stress and shear strength. A previously stable slope may be initially affected by preparatory factors, making the slope conditionally unstable. Triggering factors of a slope failure can be climatic events can then make a slope actively unstable, leading to mass movements. Mass movements can be caused by increases in shear stress, such as loading, lateral pressure, and transient forces. Alternatively, shear strength may be decreased by weathering, changes in pore water pressure, and organic material.

Several modes of failure for earth slopes include falls, topples, slides, and flows. In slopes with coarse grained soil or rocks, falls typically occur as the rapid descent of rocks and other loose slope material. A slope topples when a large column of soil tilts over its vertical axis at failure. Typical slope stability analysis considers sliding failures, categorized mainly as rotational slides or translational slides. As implied by the name, rotational slides fail along a generally curved surface, while translational slides fail along a more planar surface. A slope failing as a flow would resemble a fluid flowing downhill.

Slope Stability Analysis

Stability analysis is needed for the design of engineered slopes and for estimating the risk of slope failure in natural or designed slopes. A common assumption is that a slope consists of a layer of soil sitting on top of a rigid base. The mass and the base are assumed to interact via friction. The interface between the mass and the base can be planar, curved, or have some other complex geometry. The goal of a slope stability analysis is to determine the conditions under which the mass will slip relative to the base and lead to slope failure.

If the interface between the mass and the base of a slope has a complex geometry, slope stability analysis is difficult and numerical solution methods are required. Typically, the exact geometry of the interface is not known and a simplified interface geometry is assumed. Finite slopes require three-dimensional models to be analyzed. To keep the problem simple, most slopes are analyzed assuming that the slopes are infinitely wide and can therefore be represented by two-dimensional models. A slope can be drained or undrained. The undrained condition is used in the calculations to produce conservative estimates of risk.

A popular stability analysis approach is based on principles pertaining to the limit equilibrium concept. This method analyzes a finite or infinite slope as if it were about to fail along its sliding failure surface. Equilibrium stresses are calculated along the failure plane, and compared to the soils shear strength as determined by Terzaghi's shear strength equation. Stability is ultimately decided by a factor of safety equal to the ratio of shear strength to the equilibrium stresses along the failure surface. A factor of safety greater than one generally implies a stable slope, failure of which should not occur assuming the slope is undisturbed. A factor of safety of 1.5 for static conditions is commonly used in practice.

Offshore Geotechnical Engineering

Platforms offshore Mexico.

Offshore (or *marine*) *geotechnical engineering* is concerned with foundation design for human-made structures in the sea, away from the coastline (in opposition to *onshore* or *nearshore*). Oil platforms, artificial islands and submarine pipelines are examples of such structures. There are number of significant differences between onshore and offshore geotechnical engineering. Notably, ground improvement (on the seabed) and site investigation are more expensive, the offshore structures are exposed to a wider range of geohazards, and the environmental and financial consequences are higher in case of failure. Offshore structures are exposed to various environmental loads, notably wind, waves and currents. These phenomena may affect the integrity or the serviceability of the structure and its foundation during its operational lifespan – they need to be taken into account in offshore design.

In subsea geotechnical engineering, seabed materials are considered a two-phase material composed of 1) rock or mineral particles and 2) water. Structures may be fixed in place in the seabed—as is the case for piers, jettys and fixed-bottom wind turbines—or may be a floating structure that remain roughly fixed relative to its geotechnical anchor point. Undersea mooring of human-engineered floating structures include a large number of offshore oil and gas platforms and, since 2008, a few floating wind turbines. Two common types of engineered design for anchoring floating structures include tension-leg and catenary loose mooring systems. "Tension leg mooring systems have vertical tethers under tension providing large restoring moments in pitch and roll. Catenary mooring systems provide station keeping for an offshore structure yet provide little stiffness at low tensions."

Geosynthetics

A collage of geosynthetic products.

Geosynthetics are a type of plastic polymer products used in geotechnical engineering that improve engineering performance while reducing costs. This includes geotextiles, geogrids, geomembranes, geocells, and geocomposites. The synthetic nature of the products make them suitable for use in the ground where high levels of durability are required; their main functions include: drainage, filtration, reinforcement, separation and containment. Geosynthetics are available in a wide range of forms and materials, each to suit a slightly different end use, although they are frequently used together. These products have a wide range of applications and are currently used in many civil and geotechnical engineering applications including: roads, airfields, railroads, embankments, piled embankments, retaining structures, reservoirs, canals, dams, landfills, bank protection and coastal engineering.

REFERENCES

1. Holtz, R. and Kovacs, W. (1981), An Introduction to Geotechnical Engineering, Prentice-Hall, Inc. ISBN 0-13-484394-0
2. Bowles, J. (1988), Foundation Analysis and Design, McGraw-Hill Publishing Company. ISBN 0-07-006776-7
3. Cedergren, Harry R. (1977), Seepage, Drainage, and Flow Nets, Wiley. ISBN 0-471-14179-8
4. Kramer, Steven L. (1996), Geotechnical Earthquake Engineering, Prentice-Hall, Inc. ISBN 0-13-374943-6
5. Freeze, R.A. & Cherry, J.A., (1979), Groundwater, Prentice-Hall. ISBN 0-13-

365312-9

6. Lunne, T. & Long, M.,(2006), Review of long seabed samplers and criteria for new sampler design, Marine Geology, Vol 226, p. 145-165
7. Mitchell, James K. & Soga, K. (2005), Fundamentals of Soil Behavior 3rd ed., John Wiley & Sons, Inc. ISBN 978-0-471-46302-3
8. Rajapakse, Ruwan., (2005), "Pile Design and Construction", 2005. ISBN 0-9728657-1-3
9. Fang, H.-Y. and Daniels, J. (2005) Introductory Geotechnical Engineering : an environmental perspective, Taylor & Francis. ISBN 0-415-30402-4
10. NAVFAC (Naval Facilities Engineering Command) (1986) Design Manual 7.01, Soil Mechanics, US Government Printing Office
11. NAVFAC (Naval Facilities Engineering Command) (1986) Design Manual 7.02, Foundations and Earth Structures, US Government Printing Office
12. NAVFAC (Naval Facilities Engineering Command) (1983) Design Manual 7.03, Soil Dynamics, Deep Stabilization and Special Geotechnical Construction, US Government Printing Office
13. Terzaghi, K., Peck, R.B. and Mesri, G. (1996), Soil Mechanics in Engineering Practice 3rd Ed., John Wiley & Sons, Inc. ISBN 0-471-08658-4
14. Santamarina, J.C., Klein, K.A., & Fam, M.A. (2001), "Soils and Waves: Particulate Materials Behavior, Characterization and Process Monitoring", Wiley, ISBN 978-0-471-49058-6
15. Firuziaan, M. and Estorff, O., (2002), "Simulation of the Dynamic Behavior of Bedding-Foundation-Soil in the Time Domain", Springer Verlag.
16. Brock, 1923, The Education of a Geologist: Economic Geology, v. 18, pp. 595-597.
17. Bates and Jackson, 1980, Glossary of Geology: American Geological Institute.
18. Kiersh, 1991, The Heritage of Engineering Geology: The First Hundred Years: Geological Society of America; Centennial Special Volume 3
19. Legget, Robert F., editor, 1982, Geology under cities: Geological Society of America; Reviews in Engineering Geology, volume V, 131 pages; contains nine articles by separate authors for these cities: Washington, DC; Boston; Chicago; Edmonton; Kansas City; New Orleans; New York City; Toronto; and Twin Cities, Minnesota.
20. Legget, Robert F., and Karrow, Paul F., 1983, Handbook of geology in civil engineering: McGraw-Hill Book Company, 1,340 pages, 50 chapters, five appendices, 771 illustrations. ISBN 0-07-037061-3
21. Price, David George, *Engineering Geology: Principles and Practice*, Springer, 2008 ISBN 3-540-29249-7
22. Prof. D. Venkat Reddy, NIT-Karnataka, **Engineering Geology**, Vikas Publishers, 2010 ISBN 978-81259-19032
23. Bulletin of Engineering Geology and the Environment

Economic Geology

Chapter Contents

Economic geology deals with metal ores, fossil fuels (e.g., petroleum, natural gas, and coal), and other materials of commercial value, such as salt, gypsum, and building stone. It applies the principles and methods of various other fields of the geologic sciences, most notably geophysics, structural geology, and stratigraphy. Its chief objective is to guide the exploration for mineral resources and help determine which deposits are economically worthwhile to mine. Specialists in economic geology often assist in the extraction of the mineral commodities as well.

INTRODUCTION

Economic geology is concerned with earth materials that can be used for economic and/or industrial purposes. These materials include precious and base metals, nonmetallic minerals, construction-grade stone, petroleum minerals, coal, and water. The term commonly refers to metallic mineral deposits and mineral resources. The techniques employed by other earth science disciplines (such as geochemistry, mineralogy, geophysics, petrology and structural geology) might all be used to understand, describe, and exploit an ore deposit.

An open pit uranium mine in Namibia

Economic geology is studied and practiced by geologists. Economic geology may be of interest to other professions such as engineers, environmental scientists, and conservationists because of the far-reaching impact that extractive industries have on society, the economy, and the environment.

PURPOSE OF STUDIES

The purpose of the study of economic geology is to gain understanding of the genesis and localization of ore deposits plus the minerals associated with ore deposits. Though metals, minerals and other geologic commodities are non-renewable in human time frames, the impression of a fixed or limited stock paradigm of scarcity has always lead to human innovation resulting in a replacement commodity substituted for those commodities which become too expensive. Additionally the fixed stock of most mineral commodities is huge (e.g., copper within the earth's crust given current rates of consumption would last for more than 100 million years. Nonetheless, economic geologists continue to successfully expand and define known mineral resources.

MINERAL RESOURCES

Mineral resources are concentrations of minerals significant for current and future societal needs. Ore is classified as mineralization economically and technically feasible for extraction. Not all mineralization meets these criteria for various reasons. The specific categories of mineralization in an economic sense are:

- *Mineral occurrences* or prospects of geological interest but not necessarily economic interest

- *Mineral resources* include those potentially economically and technically feasible and those that are not
- *Ore reserves*, which must be economically and technically feasible to extract

ORE GEOLOGY

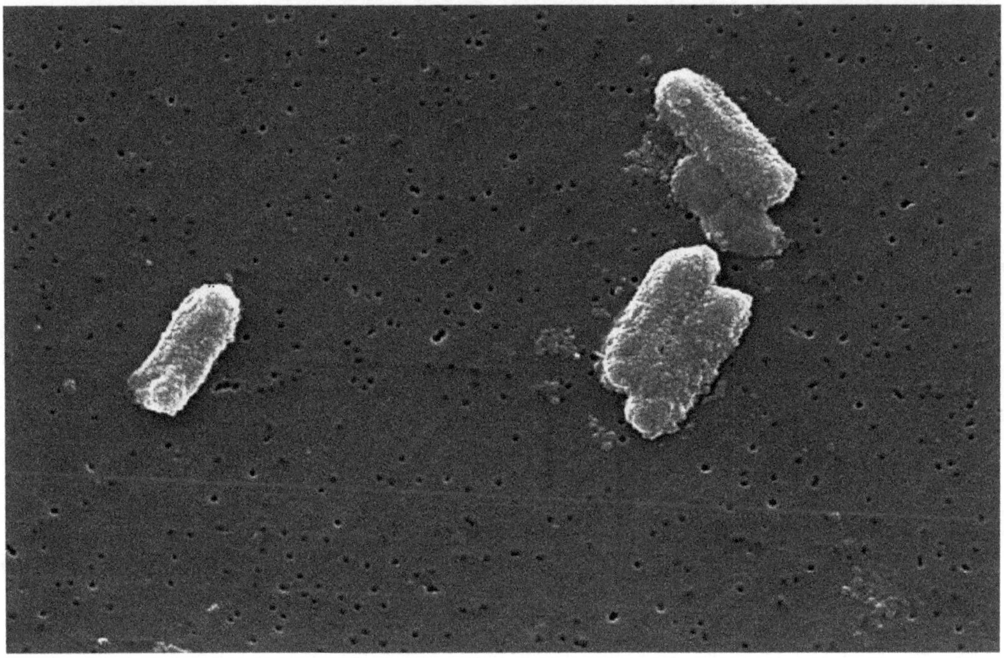

Citrobacter species can have concentrations of uranium in their bodies 300 times higher than in the surrounding environment.

Geologists are involved in the study of ore deposits, which includes the study of ore genesis and the processes within the Earth's crust that form and concentrate ore minerals into economically viable quantities.

Study of metallic ore deposits involves the use of structural geology, geochemistry, the study of metamorphism and its processes, as well as understanding metasomatism and other processes related to ore genesis.

Ore deposits are delineated by mineral exploration, which uses geochemical prospecting, drilling and resource estimation via geostatistics to quantify economic ore bodies. The ultimate aim of this process is mining.

COAL AND PETROLEUM GEOLOGY

Mud log in process, a common way to study the lithology when drilling oil wells.

The study of sedimentology is of prime importance to the delineation of economic reserves of petroleum and coal energy resources.

COAL

Coal is a combustible black or brownish-black sedimentary rock usually occurring in rock strata in layers or veins called **coal beds** or **coal seams**. The harder forms, such as anthracite coal, can be regarded as metamorphic rock because of later exposure to elevated temperature and pressure. Coal is composed primarily of carbon, along with variable quantities of other elements, chiefly hydrogen, sulfur, oxygen, and nitrogen. A fossil fuel, coal forms when dead plant matter is converted into peat, which in turn is converted into lignite, then sub-bituminous coal, after that bituminous coal, and lastly anthracite. This involves biological and geological processes that take place over time.

Throughout history, coal has been used as an energy resource, primarily burned for the production of electricity and heat, and is also used for industrial purposes, such as refining metals. Coal is the largest source of energy for the generation of electricity worldwide, as well as one of the largest worldwide anthropogenic sources of carbon dioxide releases. The extraction of coal, its use in energy production and its byproducts are all associated with environmental and health effects including climate change.

Bituminous coal

Coal is extracted from the ground by coal mining. Since 1983, the world's top coal producer has been China. In 2011 China produced 3,520 million tonnes of coal – 49.5% of 7,695 million tonnes world coal production. In 2011 other large producers were United States (993 million tonnes), India (589), European Union (576) and Australia (416). In 2010 the largest exporters were Australia with 328 million tonnes (27.1% of world coal export) and Indonesia with 316 million tonnes (26.1%), while the largest importers were Japan with 207 million tonnes (17.5% of world coal import), China with 195 million tonnes (16.6%) and South Korea with 126 million tonnes (10.7%).

Etymology

The word originally took the form *col* in Old English, from Proto-Germanic **kula*(*n*), which in turn is hypothesized to come from the Proto-Indo-European root **g*(*e*)*u-lo-* "live coal". Germanic cognates include the Old Frisian *kole*, Middle Dutch *cole*, Dutch *kool*, Old High German *chol*, German *Kohle* and Old Norse *kol*, and the Irish word *gual* is also a cognate via the Indo-European root. In Old Turkic languages, *kül* is "ash(es), cinders, *öčür* is quench. The compound charcoal in Turkic is *öčür(ülmüş) kül*, literally "quenched ashes, cinders, coals" with elided anlaut *ö-* and inflection affixes *-ülmüş*.

The word took on the meaning mineral consisting of fossilized carbon in the thirteenth century.

Formation

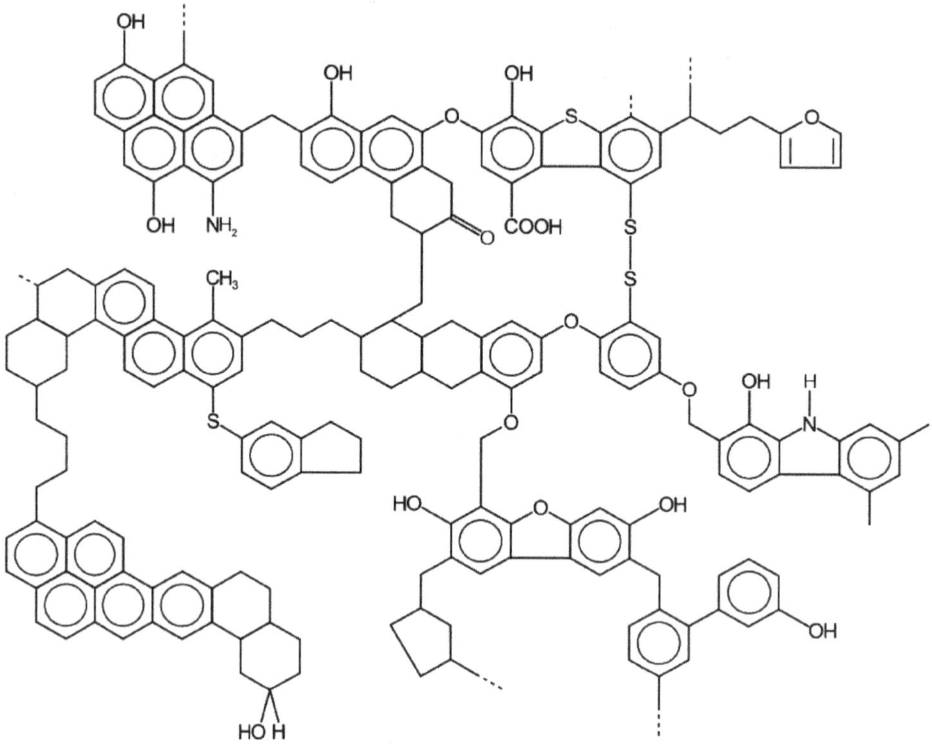

Example chemical structure of coal

At various times in the geologic past, the Earth had dense forests in low-lying wetland areas. Due to natural processes such as flooding, these forests were buried underneath soil. As more and more soil deposited over them, they were compressed. The temperature also rose as they sank deeper and deeper. As the process continued the plant matter was protected from biodegradation and oxidation, usually by mud or acidic water. This trapped the carbon in immense peat bogs that were eventually covered and deeply buried by sediments. Under high pressure and high temperature, dead vegetation was slowly converted to coal. As coal contains mainly carbon, the conversion of dead vegetation into coal is called carbonization.

The wide, shallow seas of the Carboniferous Period provided ideal conditions for coal formation, although coal is known from most geological periods. The exception is the coal gap in the Permian–Triassic extinction event, where coal is rare. Coal is known from Precambrian strata, which predate land plants — this coal is presumed to have originated from residues of algae.

Ranks

Coastal exposure of the Point Aconi Seam (Nova Scotia)

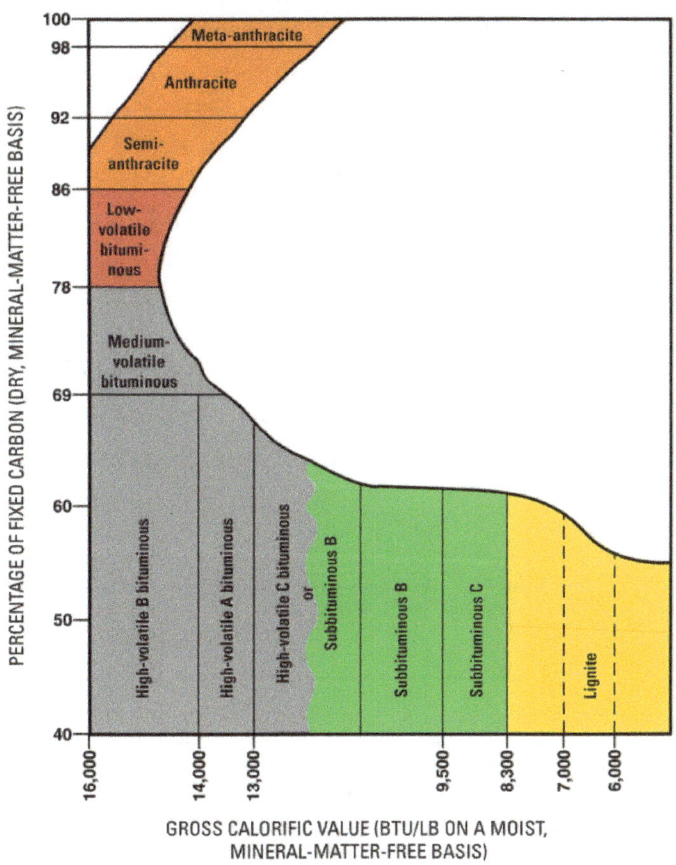

Coal ranking system used in the United States (US Geological Survey)

As geological processes apply pressure to dead biotic material over time, under suitable conditions, its metamorphic grade increases successively into:

- Peat, considered to be a precursor of coal, has industrial importance as a fuel in some regions, for example, Ireland and Finland. In its dehydrated form, peat is a highly effective absorbent for fuel and oil spills on land and water. It is also used as a conditioner for soil to make it more able to retain and slowly release water.
- Lignite, or brown coal, is the lowest rank of coal and used almost exclusively as fuel for electric power generation. Jet, a compact form of lignite, is sometimes polished and has been used as an ornamental stone since the Upper Palaeolithic.
- Sub-bituminous coal, whose properties range from those of lignite to those of bituminous coal, is used primarily as fuel for steam-electric power generation and is an important source of light aromatic hydrocarbons for the chemical synthesis industry.
- Bituminous coal is a dense sedimentary rock, usually black, but sometimes dark brown, often with well-defined bands of bright and dull material; it is used primarily as fuel in steam-electric power generation, with substantial quantities used for heat and power applications in manufacturing and to make coke.
- Steam coal is a grade between bituminous coal and anthracite, once widely used as a fuel for steam locomotives. In this specialized use, it is sometimes known as sea coal in the US. Small steam coal (dry small steam nuts or DSSN) was used as a fuel for domestic water heating.
- Anthracite, the highest rank of coal, is a harder, glossy black coal used primarily for residential and commercial space heating. It may be divided further into metamorphically altered bituminous coal and petrified oil, as from the deposits in Pennsylvania.
- Graphite is one of the more difficult coals to ignite and is not commonly used as fuel — it is mostly used in pencils, and when powdered, as a lubricant.

The classification of coal is generally based on the content of volatiles. However, the exact classification varies between countries. According to the German classification, coal is classified as follows:

German Classification	English Designation	Volatiles %	C Carbon %	H Hydrogen %	O Oxygen %	S Sulfur %	Heat content kJ/kg
Braunkohle	Lignite (brown coal)	45–65	60–75	6.0–5.8	34-17	0.5-3	<28,470
Flammkohle	Flame coal	40-45	75-82	6.0-5.8	>9.8	~1	<32,870
Gasflammkohle	Gas flame coal	35-40	82-85	5.8-5.6	9.8-7.3	~1	<33,910
Gaskohle	Gas coal	28-35	85-87.5	5.6-5.0	7.3-4.5	~1	<34,960
Fettkohle	Fat coal	19-28	87.5-89.5	5.0-4.5	4.5-3.2	~1	<35,380
Esskohle	Forge coal	14-19	89.5-90.5	4.5-4.0	3.2-2.8	~1	<35,380

Magerkohle	Nonbaking coal	10-14	90.5-91.5	4.0-3.75	2.8-3.5	~1	35,380
Anthrazit	Anthracite	7-12	>91.5	<3.75	<2.5	~1	<35,300
Note, the percentages are percent by mass of the indicated elements							

The middle six grades in the table represent a progressive transition from the English-language sub-bituminous to bituminous coal, while the last class is an approximate equivalent to anthracite, but more inclusive (US anthracite has < 6% volatiles). Cannel coal (sometimes called "candle coal") is a variety of fine-grained, high-rank coal with significant hydrogen content. It consists primarily of "exinite macerals, now termed liptinite.

Hilt's law

Hilt's law is a geological term that states that, in a small area, the deeper the coal, the higher its rank (grade). The law holds true if the thermal gradient is entirely vertical, but metamorphism may cause lateral changes of rank, irrespective of depth.

Content

Average content of some other elements	
Substance	**Content**
Mercury (Hg)	0.10±0.01 ppm
Arsenic (As)	1.4 – 71 ppm
Selenium (Se)	3 ppm

Early Uses as Fuel

Chinese coal miners in an illustration of the *Tiangong Kaiwu* encyclopedia, published in 1637

The earliest recognized use is from the Shenyang area of China 4000 BC where Neolithic inhabitants had begun carving ornaments from black lignite. Coal from the Fushun mine in northeastern China was used to smelt copper as early as 1000 BCE. Marco Polo, the Italian who traveled to China in the 13th century, described coal as black stones ... which burn like logs, and said coal was so plentiful, people could take three hot baths a week. In Europe, the earliest reference to the use of coal as fuel is from the geological treatise *On stones* (Lap. 16) by the Greek scientist Theophrastus (*circa* 371–287 BC):

Among the materials that are dug because they are useful, those known as *anthrakes* [coals] are made of earth, and, once set on fire, they burn like charcoal. They are found in Liguria ... and in Elis as one approaches Olympia by the mountain road; and they are used by those who work in metals.

—Theophrastus, On Stones (16) translation

Outcrop coal was used in Britain during the Bronze Age (3000–2000 BC), where it has been detected as forming part of the composition of funeral pyres. In Roman Britain, with the exception of two modern fields, the Romans were exploiting coals in all the major coalfields in England and Wales by the end of the second century AD". Evidence of trade in coal (dated to about AD 200) has been found at the Roman settlement at Heronbridge, near Chester, and in the Fenlands of East Anglia, where coal from the Midlands was transported via the Car Dyke for use in drying grain. Coal cinders have been found in the hearths of villas and Roman forts, particularly in Northumberland, dated to around AD 400. In the west of England, contemporary writers described the wonder of a permanent brazier of coal on the altar of Minerva at Aquae Sulis (modern day Bath), although in fact easily accessible surface coal from what became the Somerset coalfield was in common use in quite lowly dwellings locally. Evidence of coal's use for iron-working in the city during the Roman period has been found. In Eschweiler, Rhineland, deposits of bituminous coal were used by the Romans for the smelting of iron ore.

Coal miner in Britain, 1942

No evidence exists of the product being of great importance in Britain before the High

Middle Ages, after about AD 1000. Mineral coal came to be referred to as "seacoal in the 13th century; the wharf where the material arrived in London was known as Seacoal Lane, so identified in a charter of King Henry III granted in 1253. Initially, the name was given because much coal was found on the shore, having fallen from the exposed coal seams on cliffs above or washed out of underwater coal outcrops, but by the time of Henry VIII, it was understood to derive from the way it was carried to London by sea. In 1257–59, coal from Newcastle upon Tyne was shipped to London for the smiths and lime-burners building Westminster Abbey. Seacoal Lane and Newcastle Lane, where coal was unloaded at wharves along the River Fleet, are still in existence. (See Industrial processes below for modern uses of the term.)

These easily accessible sources had largely become exhausted (or could not meet the growing demand) by the 13th century, when underground extraction by shaft mining or adits was developed. The alternative name was "pitcoal, because it came from mines. The development of the Industrial Revolution led to the large-scale use of coal, as the steam engine took over from the water wheel. In 1700, five-sixths of the world›s coal was mined in Britain. Britain would have run out of suitable sites for watermills by the 1830s if coal had not been available as a source of energy. In 1947, there were some 750,000 miners in Britain, but by 2004, this had shrunk to some 5,000 miners working in around 20 collieries.

Uses Today

Castle Gate Power Plant near Helper, Utah, USA

Coal rail cars

Coal as fuel

Coal is primarily used as a solid fuel to produce electricity and heat through combustion. World coal consumption was about 7.25 billion tonnes in 2010 (7.99 billion short tons) and is expected to increase 48% to 9.05 billion tonnes (9.98 billion short tons) by 2030. China produced 3.47 billion tonnes (3.83 billion short tons) in 2011. India produced about 578 million tonnes (637.1 million short tons) in 2011. 68.7% of China›s electricity comes from coal. The US consumed about 13% of the world total in 2010, i.e. 951 million tonnes (1.05 billion short tons), using 93% of it for generation of electricity. 46% of total power generated in the US was using coal. The United States Energy Information Administration estimates coal reserves at 948×10^9 short tons (860 Gt). One estimate for resources is 18,000 Gt.

When coal is used for electricity generation, it is usually pulverized and then burned in a furnace with a boiler. The furnace heat converts boiler water to steam, which is then used to spin turbines which turn generators and create electricity. The thermodynamic efficiency of this process has been improved over time; some older coal-fired power stations have thermal efficiencies in the vicinity of 25% whereas the newest supercritical and "ultra-supercritical" steam cycle turbines, operating at temperatures over 600 °C and pressures over 27 MPa (over 3900 psi), can achieve thermal efficiencies in excess of 45% (LHV basis) using anthracite fuel, or around 43% (LHV basis) even when using lower-grade lignite fuel. Further thermal efficiency improvements are also achievable by improved pre-drying (especially relevant with high-moisture fuel such as lignite or biomass) and cooling technologies.

An alternative approach of using coal for electricity generation with improved efficiency is the integrated gasification combined cycle (IGCC) power plant. Instead of pulverizing the coal and burning it directly as fuel in the steam-generating boiler, the coal is gasified (see coal gasification) to create syngas, which is burned in a gas turbine to produce electricity (just like natural gas is burned in a turbine). Hot exhaust gases from the turbine are used to raise steam in a heat recovery steam generator which powers a supplemental steam turbine. Thermal efficiencies of current IGCC power plants range from 39% to 42% (HHV basis) or ≈42–45% (LHV basis) for bituminous coal and assuming utilization of mainstream gasification technologies (Shell, GE Gasifier, CB&I). IGCC power plants outperform conventional pulverized coal-fueled plants in terms of pollutant emissions, and allow for relatively easy carbon capture.

At least 40% of the world›s electricity comes from coal, and in 2012, about one-third of the United States› electricity came from coal, down from approximately 49% in 2008. As of 2012 in the United States, use of coal to generate electricity was declining, as plentiful supplies of natural gas obtained by hydraulic fracturing of tight shale formations became available at low prices.

In Denmark, a net electric efficiency of > 47% has been obtained at the coal-fired Nordjyllandsværket CHP Plant and an overall plant efficiency of up to 91% with cogeneration of electricity and district heating. The multifuel-fired Avedøreværket CHP Plant just outside Copenhagen can achieve a net electric efficiency as high as 49%. The overall plant efficiency with cogeneration of electricity and district heating can reach as much as 94%.

An alternative form of coal combustion is as coal-water slurry fuel (CWS), which was developed in the Soviet Union. Other ways to use coal are combined heat and power cogeneration and an MHD topping cycle.

The total known deposits recoverable by current technologies, including highly polluting, low-energy content types of coal (i.e., lignite, bituminous), is sufficient for many years. Consumption is increasing and maximal production could be reached within decades (see world coal reserves, below). On the other hand, much may have to be left in the ground to avoid climate change.

Coking Coal and Use of Coke

Coke oven at a smokeless fuel plant in Wales, United Kingdom

Coke is a solid carbonaceous residue derived from low-ash, low-sulfur bituminous coal (metallurgical coal), from which the volatile constituents are driven off by baking in an oven without oxygen at temperatures as high as 1,000 °C (1,832 °F), so the fixed carbon and residual ash are fused together. Metallurgical coke is used as a fuel and as a reducing agent in smelting iron ore in a blast furnace. The result is pig iron, and is too rich in dissolved carbon, so it must be treated further to make steel. The coking coal should be low in sulfur and phosphorus, so they do not migrate to the metal. Based on the ash percentage, the coking coal can be divided into various grades. These grades are:

- Steel Grade I (Ash content not exceeding 15%)
- Steel Grade II (Exceeding 15% but not exceeding 18%)
- Washery Grade I (Exceeding 18% but not exceeding 21%)
- Washery Grade II (Exceeding 21% but not exceeding 24%)
- Washery Grade III (Exceeding 24% but not exceeding 28%)
- Washery Grade IV (Exceeding 28% but not exceeding 35%)

The coke must be strong enough to resist the weight of overburden in the blast furnace, which is why coking coal is so important in making steel using the conventional route. However, the alternative route is direct reduced iron, where any carbonaceous fuel can be used to make sponge or pelletised iron. Coke from coal is grey, hard, and porous and has a heating value of 24.8 million Btu/ton (29.6 MJ/kg). Some cokemaking processes produce valuable byproducts, including coal tar, ammonia, light oils, and coal gas.

Petroleum coke is the solid residue obtained in oil refining, which resembles coke, but contains too many impurities to be useful in metallurgical applications.

Gasification

Coal gasification can be used to produce syngas, a mixture of carbon monoxide (CO) and hydrogen (H_2) gas. Often syngas is used to fire gas turbines to produce electricity, but the versatility of syngas also allows it to be converted into transportation fuels, such as gasoline and diesel, through the Fischer-Tropsch process; alternatively, syngas can be converted into methanol, which can be blended into fuel directly or converted to gasoline via the methanol to gasoline process. Gasification combined with Fischer-Tropsch technology is currently used by the Sasol chemical company of South Africa to make motor vehicle fuels from coal and natural gas. Alternatively, the hydrogen obtained from gasification can be used for various purposes, such as powering a hydrogen economy, making ammonia, or upgrading fossil fuels. During gasification, the coal is mixed with oxygen and steam while also being heated and pressurized. During the reaction, oxygen and water molecules oxidize the coal into carbon monoxide (CO), while also releasing hydrogen gas (H_2). This process has been conducted in both underground coal mines and in the production of town gas which was piped to customers to burn for illumination, heating, and cooking.

$$\text{C } (\textit{as Coal}) + O_2 + H_2O \rightarrow H_2 + CO$$

If the refiner wants to produce gasoline, the syngas is collected at this state and routed into a Fischer-Tropsch reaction. If hydrogen is the desired end-product, however, the syngas is fed into the water gas shift reaction, where more hydrogen is liberated.

$$CO + H_2O \rightarrow CO_2 + H_2$$

Liquefaction

Coal can also be converted into synthetic fuels equivalent to gasoline or diesel by several different direct processes (which do not intrinsically require gasification or indirect conversion). In the direct liquefaction processes, the coal is either hydrogenated or carbonized. Hydrogenation processes are the Bergius process, the SRC-I and SRC-II (Solvent Refined Coal) processes, the NUS Corporation hydrogenation processand several other single-stage and two-stage processes. In the process of low-temperature carbonization, coal is coked at temperatures between 360 and 750 °C (680 and 1,380 °F). These temperatures optimize the production of coal tars richer in lighter hydrocarbons than normal coal tar. The coal tar is then further processed into fuels. An overview of coal liquefaction and its future potential is available.

Coal liquefaction methods involve carbon dioxide (CO_2) emissions in the conversion process. If coal liquefaction is done without employing either carbon capture and storage (CCS) technologies or biomass blending, the result is lifecycle greenhouse gas footprints that are generally greater than those released in the extraction and refinement of liquid fuel production from crude oil. If CCS technologies are employed, reductions of 5–12% can be achieved in Coal to Liquid (CTL) plants and up to a 75% reduction is achievable when co-gasifying coal with commercially demonstrated levels of biomass (30% biomass by weight) in coal/biomass-to-liquids plants. For future synthetic fuel projects, carbon dioxide sequestration is proposed to avoid releasing CO_2 into the atmosphere. Sequestration adds to the cost of production.

Refined Coal

Refined coal is the product of a coal-upgrading technology that removes moisture and certain pollutants from lower-rank coals such as sub-bituminous and lignite (brown) coals. It is one form of several precombustion treatments and processes for coal that alter coal's characteristics before it is burned. The goals of precombustion coal technologies are to increase efficiency and reduce emissions when the coal is burned. Depending on the situation, precombustion technology can be used in place of or as a supplement to postcombustion technologies to control emissions from coal-fueled boilers.

Industrial Processes

Finely ground bituminous coal, known in this application as sea coal, is a constituent of foundry sand. While the molten metal is in the mould, the coal burns slowly, releasing reducing gases at pressure, and so preventing the metal from penetrating the pores of the sand. It is also contained in 'mould wash›, a paste or liquid with the same function applied to the mould before casting. Sea coal can be mixed with the clay lining (the bod) used for the bottom of a cupola furnace. When heated, the coal decomposes and the bod becomes slightly friable, easing the process of breaking open holes for tapping the molten metal.

Production of Chemicals

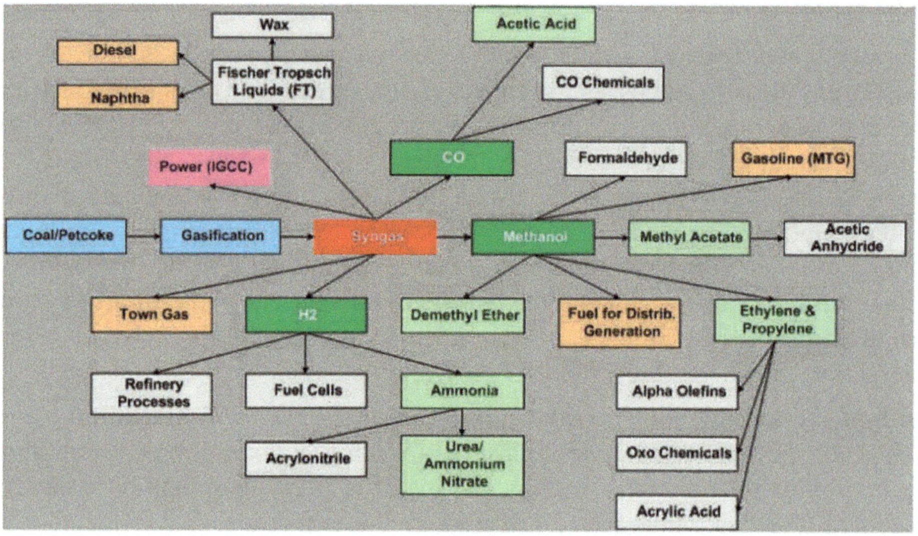

Production of chemicals from coal

Coal is an important feedstock in production of a wide range of chemical fertilizers and other chemical products. The main route to these products is coal gasification to produce syngas. Primary chemicals that are produced directly from the syngas include methanol, hydrogen and carbon monoxide, which are the chemical building blocks from which a whole spectrum of derivative chemicals are manufactured, including olefins, acetic acid, formaldehyde, ammonia, urea and others. The versatility of syngas as a precursor to primary chemicals and high-value derivative products provides the option of using relatively inexpensive coal to produce a wide range of valuable commodities.

Historically, production of chemicals from coal has been used since the 1950s and has become established in the market. According to the 2010 Worldwide Gasification Database, a survey of current and planned gasifiers, from 2004 to 2007 chemical production increased its gasification product share from 37% to 45%. From 2008 to 2010, 22% of new gasifier additions were to be for chemical production.

Because the slate of chemical products that can be made via coal gasification can in general also use feedstocks derived from natural gas and petroleum, the chemical industry tends to use whatever feedstocks are most cost-effective. Therefore, interest in using coal tends to increase for higher oil and natural gas prices and during periods of high global economic growth that may strain oil and gas production. Also, production of chemicals from coal is of much higher interest in countries like South Africa, China, India and the United States where there are abundant coal resources. The abundance of coal combined with lack of natural gas resources in China is strong inducement for the coal to chemicals industry pursued there. In the United States, the best example of the industry is Eastman Chemical Company which has been successfully operating a coal-to-chemicals plant at its Kingsport, Tennessee, site since 1983. Similarly, Sasol has built and operated coal-to-chemicals facilities in South Africa.

Coal to chemical processes do require substantial quantities of water. As of 2013 much of the coal to chemical production was in the People›s Republic of China where environmental regulation and water management was weak.

Coal Industry

Coal as a Traded Commodity

In North America, Central Appalachian coal futures contracts are currently traded on the New York Mercantile Exchange (trading symbol *QL*). The trading unit is 1,550 short tons (1,410 t) per contract, and is quoted in U.S. dollars and cents per ton. Since coal is the principal fuel for generating electricity in the United States, coal futures contracts provide coal producers and the electric power industry an important tool for hedging and risk management.

In addition to the NYMEX contract, the Intercontinental Exchange (ICE) has European (Rotterdam) and South African (Richards Bay) coal futures available for trading. The trading unit for these contracts is 5,000 tonnes (5,500 short tons), and are also quoted in U.S. dollars and cents per ton.

The price of coal increased from around $30.00 per short ton in 2000 to around $150.00 per short ton as of September 2008. As of October 2008, the price per short ton had declined to $111.50. Prices further declined to $71.25 as of October 2010. In early 2015, it was trading near $56/ton.

Environmental and Health Effects

Aerial photograph of Kingston Fossil Plant coal fly ash slurry spill site taken the day after the event

Health Effects

The use of coal as fuel causes adverse health impacts and deaths.

The deadly London smog was caused primarily by the heavy use of coal. In the United States coal-fired power plants were estimated in 2004 to cause nearly 24,000 premature deaths every year, including 2,800 from lung cancer. Annual health costs in Europe from use of coal to generate electricity are €42.8 billion, or $55 billion. Yet the disease and mortality burden of coal use today falls most heavily upon China.

Breathing in coal dust causes coalworker›s pneumoconiosis which is known colloquially as "black lung", so-called because the coal dust literally turns the lungs black from their usual pink color. In the United States alone, it is estimated that 1500 former employees of the coal industry die every year from the effects of breathing in coal mine dust.

Around 10% of coal is ash, Coal ash is hazardous and toxic to human beings and other living things.Coal ash contains the radioactive elements uranium and thorium. Coal ash and other solid combustion byproducts are stored locally and escape in various ways that expose those living near coal plants to radiation and environmental toxics.

Huge amounts of coal ash and other waste is produced annually. In 2013, the US alone consumed on the order of 983 million short tonnes of coal per year. Use of coal on this scale generates hundreds of millions of tons of ash and other waste products every year. These include fly ash, bottom ash, and flue-gas desulfurization sludge, that contain mercury, uranium, thorium, arsenic, and other heavy metals, along with non-metals such as selenium.

The American Lung Association, the American Nurses› Association, and the Physicians for Social Responsibility released a report in 2009 which details in depth the detrimental impact of the coal industry on human health, including workers in the mines and individuals living in communities near plants burning coal as a power source. This report provides medical information regarding damage to the lungs, heart, and nervous system of Americans caused by the burning of coal as fuel. It details how the air pollution caused by the plume of coal smokestack emissions is a cause of asthma, strokes, reduced intelligence, artery blockages, heart attacks, congestive heart failure, cardiac arrhythmias, mercury poisoning, arterial occlusion, and lung cancer.

More recently, the Chicago School of Public Health released a largely similar report, echoing many of the same findings.

Though coal burning has increasingly been supplanted by less-toxic natural gas use in recent years, a 2010 study by the Clean Air Task Force still estimated that air pollution from coal-fired power plants accounts for more than 13,000 premature deaths, 20,000 heart attacks, and 1.6 million lost workdays in the U.S. each year. The total monetary cost of these health impacts is over $100 billion annually.

The WHO classifies coal as a dirty fuel and encourages the movement away from such fuels towards cleaner alternatives.

Environmental Effects

Coal mining and coal fueling of power station and industrial processes can cause major environmental damage.

Water systems are affected by coal mining. For example, mining affects groundwater and water table levels and acidity. Spills of fly ash, such as the Kingston Fossil Plant coal fly ash slurry spill, can also contaminate land and waterways, and destroy homes. Power stations that burn coal also consume large quantities of water. This can affect the flows of rivers, and has consequential impacts on other land uses.

One of the earliest known impacts of coal on the water cycle was acid rain. Approximately 75 Tg/S per year of sulfur dioxide (SO_2) is released from burning coal. After release, the sulfur dioxide is oxidized to gaseous H_2SO_2 which scatters solar radiation, hence its increase in the atmosphere exerts a cooling effect on climate. This beneficially masks some of the warming caused by increased greenhouse gases. However, the sulfur is precipitated out of the atmosphere as acid rain in a matter of weeks, whereas carbon dioxide remains in the atmosphere for hundreds of years. Release of SO_2 also contributes to the widespread acidification of ecosystems.

Disused coal mines can also cause issues. Subsidence can occur above tunnels, causing damage to infrastructure or cropland. Coal mining can also cause long lasting fires, and it has been estimated that thousands of coal seam fires are burning at any given time. For example, there is a coal seam fire in Germany that has been burning since 1668, and is still burning in the 21st century.

Some environmental impacts are modest, such as dust nuisance. However, perhaps the largest and most long term effect of coal use is the release of carbon dioxide, a greenhouse gas that causes climate change and global warming, according to the IPCC and the EPA. Coal is the largest contributor to the human-made increase of CO_2 in the atmosphere.

The production of coke from coal produces ammonia, coal tar, and gaseous compounds as by-products which if discharged to land, air or waterways can act as environmental pollutants. The Whyalla steelworks is one example of a coke producing facility where liquid ammonia is discharged to the marine environment.

In 1999, world gross carbon dioxide emissions from coal usage were 8,666 million tonnes of carbon dioxide. In 2011, world gross emissions from coal usage were 14,416 million tonnes. For every megawatt-hour generated, coal-fired electric power generation emits around 2,000 pounds of carbon dioxide, which is almost double the approximately 1100 pounds of carbon dioxide released by a natural gas-fired electric plant. Because of this higher carbon efficiency of natural gas generation, as the market in the United States has changed to reduce coal and increase natural gas generation, carbon dioxide emissions may have fallen. Those measured in the first quarter of 2012 were the lowest of any recorded for the first quarter of any year since 1992. In 2013, the head of the UN climate agency advised that most of the world›s coal reserves should be left in the ground to avoid catastrophic global warming.

Clean Coal Technology

"Clean" coal technology is a collection of technologies being developed to mitigate the environmental impact of coal energy generation. Those technologies are being developed to remove or reduce pollutant emissions to the atmosphere. Some of the techniques that would be used to accomplish this include chemically washing minerals and impurities from the coal, gasification (see also IGCC), improved technology for treating flue gases to remove pollutants to increasingly stringent levels and at higher efficiency, carbon capture and storage technologies to capture the carbon dioxide from the flue gas and dewatering lower rank coals (brown coals) to improve the calorific value, and thus the efficiency of the conversion into electricity. Figures from the United States Environmental Protection Agency show that these technologies have made today›s coal-based generating fleet 77 percent cleaner on the basis of regulated emissions per unit of energy produced.

Clean coal technology usually addresses atmospheric problems resulting from burning coal. Historically, the primary focus was on SO_2 and NO_x, the most important gases in causation of acid rain, and particulates which cause visible air pollution and deleterious effects on human health. More recent focus has been on carbon dioxide (due to its impact on global warming) and concern over toxic species such as mercury. Concerns exist regarding the economic viability of these technologies and the timeframe of delivery, potentially high hidden economic costs in terms of social and environmental damage, and the costs and viability of disposing of removed carbon and other toxic matter.

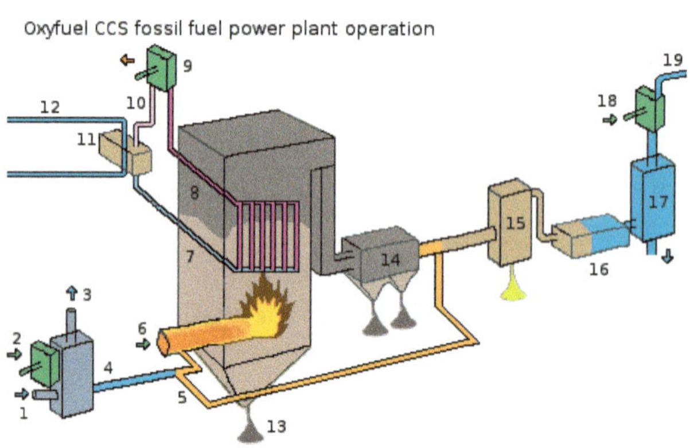

An oxyfuel CCS power plant operation processes the exhaust gases so as to separate the CO_2 so that it may be stored or sequestered

Several different technological methods are available for the purpose of carbon capture as demanded by the clean coal concept:

- Pre-combustion capture – This involves gasification of a feedstock (such as coal) to form synthesis gas, which may be shifted to produce a H_2 and CO_2-rich gas mixture, from which the CO_2 can be efficiently captured and separated, transported, and ultimately sequestered, This technology is usually associated with Integrated Gasification Combined Cycle process configurations.

- Post-combustion capture – This refers to capture of CO_2 from exhaust gases of combustion processes, typically using sorbents, solvents, or membrane separations to remove CO_2 from the bulk gases.
- Oxy-fuel combustion – Fossil fuels such as coal are burned in a mixture of recirculated flue gas and oxygen, rather than in air, which largely eliminates nitrogen from the flue gas enabling efficient, low-cost CO_2 capture.

The Kemper County IGCC Project, a 582 MW coal gasification-based power plant, will use pre-combustion capture of CO_2 to capture 65% of the CO_2 the plant produces, which will be utilized/geologically sequestered in enhanced oil recovery operations.

The Saskatchewan Government›s Boundary Dam Integrated Carbon Capture and Sequestration Demonstration Project will use post-combustion, amine-based scrubber technology to capture 90% of the CO_2 emitted by Unit 3 of the power plant; this CO_2 will be pipelined to and utilized for enhanced oil recovery in the Weyburn oil fields.

An early example of a coal-based plant using (oxy-fuel) carbon-capture technology is Swedish company Vattenfall›s Schwarze Pumpe power station located in Spremberg, Germany, built by German firm Siemens, which went on-line in September 2008. The facility captures $CO_2$and acid rain producing pollutants, separates them, and compresses the CO_2 into a liquid. Plans are to inject the CO_2 into depleted natural gas fields or other geological formations. Vattenfall opines that this technology is considered not to be a final solution for CO_2 reduction in the atmosphere, but provides an achievable solution in the near term while more desirable alternative solutions to power generation can be made economically practical.

Bioremediation

The white rot fungus trametes versicolor can grow on and metabolize naturally occurring coal. The bacteria diplococcus has been found to degrade coal, raising its temperature.

Economic Aspects

Coal (by liquefaction technology) is one of the backstop resources that could limit escalation of oil prices and mitigate the effects of transportation energy shortage that will occur under peak oil. This is contingent on liquefaction production capacity becoming large enough to satiate the very large and growing demand for petroleum. Estimates of the cost of producing liquid fuels from coal suggest that domestic U.S. production of fuel from coal becomes cost-competitive with oil priced at around $35 per barrel, with the $35 being the break-even cost. With oil prices as low as around $40 per barrel in the U.S. as of December 2008, liquid coal lost some of its economic allure in the U.S., but will probably be re-vitalized, similar to oil sand projects, with an oil price around $70 per barrel.

In China, due to an increasing need for liquid energy in the transportation sector, coal liquefaction projects were given high priority even during periods of oil prices below $40 per barrel. This is probably because China prefers not to be dependent on foreign oil, instead utilizing its enormous domestic coal reserves. As oil prices were increasing during the first half of 2009, the coal liquefaction projects in China were again boosted, and these projects are profitable with an oil barrel price of $40.

China is the largest producer of coal in the world. It is the world›s largest energy consumer, and relies on coal to supply 69% of its energy needs. An estimated 5 million people worked in China›s coal-mining industry in 2007.

Coal pollution costs the EU €43 billion each year. Measures to cut air pollution may have beneficial long-term economic impacts for individuals.

Energy Density and Carbon Impact

The energy density of coal, i.e. its heating value, is roughly 24 megajoules per kilogram (approximately 6.7 kilowatt-hours per kg). For a coal power plant with a 40% efficiency, it takes an estimated 325 kg (717 lb) of coal to power a 100 W lightbulb for one year.

As of 2006, the average efficiency of electricity-generating power stations was 31%; in 2002, coal represented about 23% of total global energy supply, an equivalent of 3.4 billion tonnes of coal, of which 2.8 billion tonnes were used for electricity generation.

The US Energy Information Agency›s 1999 report on CO_2 emissions for energy generation quotes an emission factor of 0.963 kg CO_2/kWh for coal power, compared to 0.881 kg CO_2/kWh (oil), or 0.569 kg CO_2/kWh (natural gas).

Underground Fires

Thousands of coal fires are burning around the world. Those burning underground can be difficult to locate and many cannot be extinguished. Fires can cause the ground above to subside, their combustion gases are dangerous to life, and breaking out to the surface can initiate surface wildfires. Coal seams can be set on fire by spontaneous combustion or contact with a mine fire or surface fire. Lightning strikes are an important source of ignition. The coal continues to burn slowly back into the seam until oxygen (air) can no longer reach the flame front. A grass fire in a coal area can set dozens of coal seams on fire. Coal fires in China burn an estimated 120 million tons of coal a year, emitting 360 million metric tons of CO_2, amounting to 2–3% of the annual worldwide production of CO_2 from fossil fuels. In Centralia, Pennsylvania (a borough located in the Coal Region of the United States), an exposed vein of anthracite ignited in 1962 due to a trash fire in the borough landfill, located in an abandoned anthracite strip mine pit. Attempts to extinguish the fire were unsuccessful, and it continues to burn underground to this day. The Australian Burning Mountain was originally believed to be a volcano, but the smoke and ash come from a coal fire that has been burning for some 6,000 years.

At Kuh i Malik in Yagnob Valley, Tajikistan, coal deposits have been burning for thousands of years, creating vast underground labyrinths full of unique minerals, some of them very beautiful. Local people once used this methodto mine ammoniac. This place has been well-known since the time of Herodotus, but European geographers misinterpreted the Ancient Greek descriptions as the evidence of active volcanism in Turkestan (up to the 19th century, when the Russian army invaded the area.

The reddish siltstone rock that caps many ridges and buttes in the Powder River Basin in Wyoming and in western North Dakota is called *porcelanite*, which resembles the coal burning waste "clinker" or volcanic "scoria". Clinker is rock that has been fused by the natural burning of coal. In the Powder River Basin approximately 27 to 54 billion tons of coal burned within the past three million years. Wild coal fires in the area were reported by the Lewis and Clark Expedition as well as explorers and settlers in the area.

Production Trends

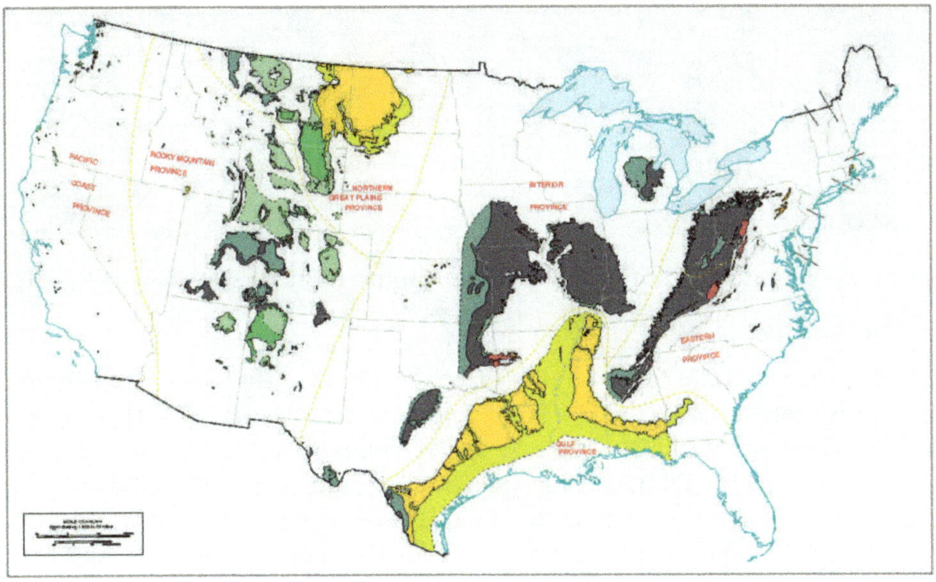

Continental United States coal regions

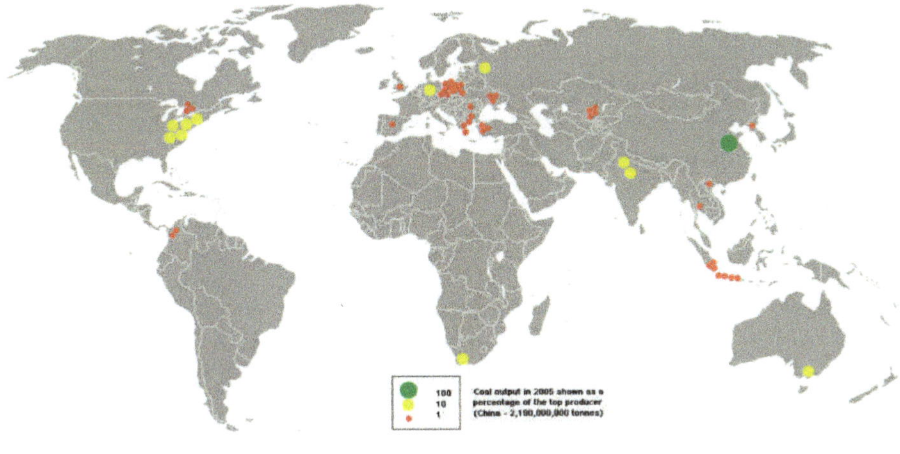

Coal output in 2005

A coal mine in Wyoming, United States. The United States has the world's largest coal reserves.

In 2006, China was the top producer of coal with 38% share followed by the United States and India, according to the British Geological Survey. As of 2012 coal production in the United States was falling at the rate of 7% annually with many power plants using coal shut down or converted to natural gas; however, some of the reduced domestic demand was taken up by increased exports with five coal export terminals being proposed in the Pacific Northwest to export coal from the Powder River Basin to China and other Asian markets; however, as of 2013, environmental opposition was increasing. High-sulfur coal mined in Illinois which was unsaleable in the United States found a ready market in Asia as exports reached 13 million tons in 2012.

World Coal Reserves

The 948 billion short tons of recoverable coal reserves estimated by the Energy Information Administration are equal to about 4,196 BBOE (billion barrels of oil equivalent). The amount of coal burned during 2007 was estimated at 7.075 billion short tons, or 133.179 quadrillion BTU's. This is an average of 18.8 million BTU per short ton. In terms of heat content, this is about 57,000,000 barrels (9,100,000 m^3) of oil equivalent per day. By comparison in 2007, natural gas provided 51,000,000 barrels (8,100,000 m^3) of oil equivalent per day, while oil provided 85,800,000 barrels (13,640,000 m^3) per day.

British Petroleum, in its 2007 report, estimated at 2006 end that there were 147 years reserves-to-production ratio based on *proven* coal reserves worldwide. This figure only includes reserves classified as "proven"; exploration drilling programs by mining companies, particularly in under-explored areas, are continually providing new reserves. In many cases, companies are aware of coal deposits that have not been sufficiently drilled to qualify as "proven". However, some nations haven't updated their information and assume reserves remain at the same levels even with withdrawals.

Of the three fossil fuels, coal has the most widely distributed reserves; coal is mined in over 100 countries, and on all continents except Antarctica. The largest reserves are found in the United States, Russia, China, Australia and India. Note the table below.

Proved recoverable coal reserves at end-2008 or 2011 (million tonnes)

Country	Anthracite & Bituminous	SubBituminous	Lignite	Total	Percentage of World Total	Year
United States	108,501	98,618	30,176	237,295	22.6	2011
Russia	49,088	97,472	10,450	157,010	14.4	2011
China	62,200	33,700	18,600	114,500	12.6	2011
Australia	37,100	2,100	37,200	76,400	8.9	2011
India	56,100	0	4,500	60,600	7.0	2011
Germany	99	0	40,600	40,699	4.7	
Ukraine	15,351	16,577	1,945	33,873	3.9	
Kazakhstan	21,500	0	12,100	33,600	3.9	
South Africa	30,156	0	0	30,156	3.5	
Serbia	9	361	13,400	13,770	1.6	
Colombia	6,366	380	0	6,746	0.8	
Canada	3,474	872	2,236	6,528	0.8	
Poland	4,338	0	1,371	5,709	0.7	
Indonesia	1,520	2,904	1,105	5,529	0.6	
Brazil	0	4,559	0	4,559	0.5	
Greece	0	0	3,020	3,020	0.4	
Bosnia and Herzegovina	484	0	2,369	2,853	0.3	
Mongolia	1,170	0	1,350	2,520	0.3	
Bulgaria	2	190	2,174	2,366	0.3	
Pakistan	0	166	1,904	2,070	0.3	
Turkey	529	0	1,814	2,343	0.3	
Uzbekistan	47	0	1,853	1,900	0.2	
Hungary	13	439	1,208	1,660	0.2	
Thailand	0	0	1,239	1,239	0.1	
Mexico	860	300	51	1,211	0.1	
Iran	1,203	0	0	1,203	0.1	
Czech Republic	192	0	908	1,100	0.1	
Kyrgyzstan	0	0	812	812	0.1	
Albania	0	0	794	794	0.1	
North Korea	300	300	0	600	0.1	
New Zealand	33	205	333-7,000	571–15,000	0.1	

Proved recoverable coal reserves at end-2008 or 2011 (million tonnes)						
Country	**Anthracite & Bituminous**	**SubBituminous**	**Lignite**	**Total**	**Percentage of World Total**	**Year**
Spain	200	300	30	530	0.1	
Laos	4	0	499	503	0.1	
Zimbabwe	502	0	0	502	0.1	
Argentina	0	0	550	550	0.1	2011
All others	3,421	1,346	846	5,613	0.7	
World Total	**403,197**	**287,333**	**201,000**	**891,530**	**100**	**2011**

Major Coal Producers

The reserve life is an estimate based only on current production levels and proved reserves level for the countries shown, and makes no assumptions of future production or even current production trends. Countries with annual production higher than 100 million tonnes are shown. For comparison, data for the European Union is also shown. Shares are based on data expressed in tonnes oil equivalent.

Production of Coal by Country and year (million tonnes)											
Country	**2003**	**2004**	**2005**	**2006**	**2007**	**2008**	**2009**	**2010**	**2011**	**Share**	**Reserve Life (years)**
China	1834.9	2122.6	2349.5	2528.6	2691.6	2802.0	2973.0	3235.0	3520.0	49.5%	35
United States	972.3	1008.9	1026.5	1054.8	1040.2	1063.0	975.2	983.7	992.8	14.1%	239
India	375.4	407.7	428.4	449.2	478.4	515.9	556.0	573.8	588.5	5.6%	103
European Union	637.2	627.6	607.4	595.1	592.3	563.6	538.4	535.7	576.1	4.2%	97
Australia	350.4	364.3	375.4	382.2	392.7	399.2	413.2	424.0	415.5	5.8%	184
Russia	276.7	281.7	298.3	309.9	313.5	328.6	301.3	321.6	333.5	4.0%	471
Indonesia	114.3	132.4	152.7	193.8	216.9	240.2	256.2	275.2	324.9	5.1%	17
South Africa	237.9	243.4	244.4	244.8	247.7	252.6	250.6	254.3	255.1	3.6%	118
Germany	204.9	207.8	202.8	197.1	201.9	192.4	183.7	182.3	188.6	1.1%	216
Poland	163.8	162.4	159.5	156.1	145.9	144.0	135.2	133.2	139.2	1.4%	41
Kazakhstan	84.9	86.9	86.6	96.2	97.8	111.1	100.9	110.9	115.9	1.5%	290
World Total	**5,301.3**	**5,716.0**	**6,035.3**	**6,342.0**	**6,573.3**	**6,795.0**	**6,880.8**	**7,254.6**	**7,695.4**	**100%**	**112**

Major Coal Consumers

Countries with annual consumption higher than 20 million tonnes are shown.

Consumption of Coal by Country and year (million tonnes)					
Country	**2008**	**2009**	**2010**	**2011**	**Share**
China	2,691	2,892	3,352	3,677	50.7%

Consumption of Coal by Country and year (million tonnes)					
Country	**2008**	**2009**	**2010**	**2011**	**Share**
United States	1,017	904	951	910	12.5%
India	582	640	655	715	9.9%
Russia	227	185	232	238	3.3%
Germany	243	225	232	232	3.3%
South Africa	195	185	187	291	2.6%
Japan	185	164	187	183	2.5%
Poland	135	137	135	147	2.0%
World Total	**6,647**	**6,639**	**7,252**	**N/A**	**100%**

Major coal Exporters

Countries with annual gross export higher than 10 million tonnes are shown. In terms of net export the largest exporters are still Australia (328.1 millions tonnes), Indonesia (316.2) and Russia (100.2).

Exports of Coal by Country and year (million short tons)											
Country	**2003**	**2004**	**2005**	**2006**	**2007**	**2008**	**2009**	**2010**	**2011**	**2012**	**Share**
Indonesia	107.8	131.4	142.0	192.2	221.9	228.2	261.4	316.2	331.4	421.8	29.8%
Australia	238.1	247.6	255.0	255.0	268.5	278.0	288.5	328.1	313.6	332.4	23.5%
Russia	41.0	55.7	98.6	103.4	112.2	115.4	130.9	122.1	140.1	150.7	10.7%
United States	43.0	48.0	51.7	51.2	60.6	83.5	60.4	83.2	108.2	126.7	8.7%
Colombia	50.4	56.4	59.2	68.3	74.5	74.7	75.7	76.4	89.0	92.2	6.5%
South Africa	78.7	74.9	78.8	75.8	72.6	68.2	73.8	76.7	75.8	82.0	5.8%
Canada	27.7	28.8	31.2	31.2	33.4	36.5	31.9	36.9	37.6	38.8	2.7%
Kazakhstan	30.3	27.4	28.3	30.5	32.8	47.6	33.0	36.3	33.5	35.2	2.5%
Mongolia	0.5	1.7	2.3	2.5	3.4	4.4	7.7	18.3	26.1	24.3	1.7%
Vietnam	6.9	11.7	19.8	23.5	35.1	21.3	28.2	24.7	19.7	21.1	1.5%
China	103.4	95.5	93.1	85.6	75.4	68.8	25.2	22.7	27.5	15.2	1.1%
Poland	28.0	27.5	26.5	25.4	20.1	16.1	14.6	18.1	15.0	14.9	1.0%
Total World	**713.9**	**764.0**	**936.0**	**1,000.6**	**1,073.4**	**1,087.3**	**1,090.8**	**1,212.8**	**1,286.7**	**1,413.9**	**100%**

Major coal Importers

Countries with annual gross import higher than 20 million tonnes are shown. In terms of net import the largest importers are still Japan (206.0 millions tonnes), China (172.4) and South Korea (125.8).

Imports of Coal by Country and year (million short tons)						
Country	**2006**	**2007**	**2008**	**2009**	**2010**	**Share**
Japan	199.7	209.0	206.0	182.1	206.7	17.5%
China	42.0	56.2	44.5	151.9	195.1	16.6%
South Korea	84.1	94.1	107.1	109.9	125.8	10.7%
India	52.7	29.6	70.9	76.7	101.6	8.6%
Taiwan	69.1	72.5	70.9	64.6	71.1	6.0%
Germany	50.6	56.2	55.7	45.9	55.1	4.7%
Turkey	22.9	25.8	21.7	22.7	30.0	2.5%
United Kingdom	56.8	48.9	49.2	42.2	29.3	2.5%
Italy	27.9	28.0	27.9	20.9	23.7	1.9%
Netherlands	25.7	29.3	23.5	22.1	22.8	1.9%
Russia	28.8	26.3	34.6	26.8	21.8	1.9%
France	24.1	22.1	24.9	18.3	20.8	1.8%
United States	40.3	38.8	37.8	23.1	20.6	1.8%
Total	**991.8**	**1,056.5**	**1,063.2**	**1,039.8**	**1,178.1**	**100%**

Cultural Usage

Coal is the official state mineral of Kentucky. and the official state rock of Utah; both U.S. states have a historic link to coal mining.

Some cultures hold that children who misbehave will receive only a lump of coal from Santa Claus for Christmas in their christmas stockings instead of presents.

It is also customary and considered lucky in Scotland and the North of England to give coal as a gift on New Year›s Day. This occurs as part of First-Footing and represents warmth for the year to come.

PETROLEUM GEOLOGY

Petroleum geology is the study of origin, occurrence, movement, accumulation, and exploration of hydrocarbon fuels. It refers to the specific set of geological disciplines that are applied to the search for hydrocarbons (oil exploration).

Sedimentary Basin Analysis

Petroleum geology is principally concerned with the evaluation of seven key elements in sedimentary basins:

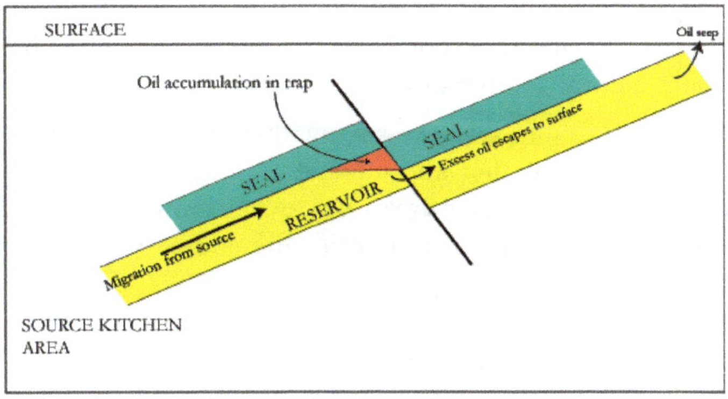

A structural trap, where a fault has juxtaposed a porous and permeable reservoir against an impermeable seal. Oil (shown in red) accumulates against the seal, to the depth of the base of the seal. Any further oil migrating in from the source will escape to the surface and seep.

- Source
- Reservoir
- Seal
- Trap
- Timing
- Maturation
- Migration

In general, all these elements must be assessed via a limited 'window' into the subsurface world, provided by one (or possibly more) exploration wells. These wells present only a 1-dimensional segment through the Earth and the skill of inferring 3-dimensional characteristics from them is one of the most fundamental in petroleum geology. Recently, the availability of inexpensive, high quality 3D seismic data (from reflection seismology) and data from various electromagnetic geophysical techniques (such as Magnetotellurics) has greatly aided the accuracy of such interpretation. The following section discusses these elements in brief. For a more in-depth treatise, see the second half of this article below.

Evaluation of the **source** uses the methods of geochemistry to quantify the nature of organic-rich rocks which contain the precursors to hydrocarbons, such that the type and quality of expelled hydrocarbon can be assessed.

The **reservoir** is a porous and permeable lithological unit or set of units that holds the hydrocarbon reserves. Analysis of reservoirs at the simplest level requires an assessment of their porosity (to calculate the volume of *in situ* hydrocarbons) and their permeability (to calculate how easily hydrocarbons will flow out of them). Some of the key disciplines used in reservoir analysis are the fields of structural analysis, stratigraphy, sedimentology, and reservoir engineering.

The **seal**, or *cap* rock, is a unit with low permeability that impedes the escape of hydrocarbons from the reservoir rock. Common seals include evaporites, chalks and shales. Analysis of seals involves assessment of their thickness and extent, such that their effectiveness can be quantified.

The **trap** is the stratigraphic or structural feature that ensures the juxtaposition of reservoir and seal such that hydrocarbons remain trapped in the subsurface, rather than escaping (due to their natural buoyancy) and being lost.

Analysis of **maturation** involves assessing the thermal history of the source rock in order to make predictions of the amount and timing of hydrocarbon generation and expulsion. Finally, careful studies of **migration** reveal information on how hydrocarbons move from source to reservoir and help quantify the source (or *kitchen*) of hydrocarbons in a particular area.

Major Subdisciplines in Petroleum Geology

Several major subdisciplines exist in petroleum geology specifically to study the seven key elements discussed above.

Source rock Analysis

In terms of source rock analysis,several facts need to be established. Firstly, the question of whether there actually *is* any source rock in the area must be answered. Delineation and identification of potential source rocks depends on studies of the local stratigraphy, palaeogeography and sedimentology to determine the likelihood of organic-rich sediments having been deposited in the past.

If the likelihood of there being a source rock is thought to be high, the next matter to address is the state of thermal maturity of the source, and the timing of maturation. Maturation of source rocks (see diagenesis and fossil fuels) depends strongly on temperature, such that the majority of oil generation occurs in the 60° to 120°C range. Gas generation starts at similar temperatures, but may continue up beyond this range, perhaps as high as 200°C. In order to determine the likelihood of oil/gas generation, therefore, the thermal history of the source rock must be calculated. This is performed with a combination of geochemical analysis of the source rock (to determine the type of kerogens present and their maturation characteristics) and basin modelling methods, such as back-stripping, to model the thermal gradient in the sedimentary column.

Basin Analysis

A full scale basin analysis is usually carried out prior to defining leads and prospects for future drilling. This study tackles the petroleum system and studies source rock (presence and quality); burial history; maturation (timing and volumes); migration and focus; and potential regional seals and major reservoir units (that define carrier beds). All these elements are used to investigate where potential hydrocarbons might migrate towards. Traps and potential leads and prospects are then defined in the area that is likely to have received hydrocarbons.

Exploration stage

Although a basin analysis is usually part of the first study a company conducts prior to moving into an area for future exploration, it is also sometimes conducted during the exploration phase. Exploration geology comprises all the activities and studies necessary for finding new hydrocarbon occurrence. Usually seismic (or 3D seismic) studies are shot, and old exploration data (seismic lines, well logs, reports) are used to expand upon the new studies. Sometimes gravity and magnetic studies are conducted, and oil seeps and spills are mapped to find potential areas for hydrocarbon occurrences. As soon as a significant hydrocarbon occurrence is found by an exploration- or wildcat-well the appraisal stage starts.

Appraisal stage

The Appraisal stage is used to delineate the extent of the discovery. Hydrocarbon reservoir properties, connectivity, hydrocarbon type and gas-oil and oil-water contacts are determined to calculate potential recoverable volumes. This is usually done by drilling more appraisal wells around the initial exploration well. Production tests may also give insight in reservoir pressures and connectivity. Geochemical and petrophysical analysis gives information on the type (viscosity, chemistry, API, carbon content, etc.) of the hydrocarbon and the nature of the reservoir (porosity, permeability, etc.).

Production Stage

After a hydrocarbon occurrence has been discovered and appraisal has indicated it is a commercial find the production stage is initiated. This stage focuses on extracting the hydrocarbons in a controlled way (without damaging the formation, within commercial favorable volumes, etc.). Production wells are drilled and completed in strategic positions. 3D seismic is usually available by this stage to target wells precisely for optimal recovery. Sometimes enhanced recovery (steam injection, pumps, etc.) is used to extract more hydrocarbons or to redevelop abandoned fields.

Reservoir Analysis

The existence of a reservoir rock (typically, sandstones and fractured limestones) is determined through a combination of regional studies (i.e. analysis of other wells in the area), stratigraphy and sedimentology (to quantify the pattern and extent of sedimentation) and seismic interpretation. Once a possible hydrocarbon reservoir is identified, the key physical characteristics of a reservoir that are of interest to a hydrocarbon explorationist are its bulk rock volume, net-to-gross ratio, porosity and permeability.

Bulk rock volume, or the gross rock volume of rock above any hydrocarbon-water contact, is determined by mapping and correlating sedimentary packages. The net-to-gross ratio, typically estimated from analogues and wireline logs, is used to calculate the proportion of the sedimentary packages that contains reservoir rocks. The bulk rock volume multiplied by the net-to-gross ratio gives the net rock volume of the reservoir. The net rock volume multiplied by porosity gives the total hydrocarbon pore volume i.e. the volume within the sedimentary package that fluids (importantly, hydrocarbons and water) can occupy. The summation of these volumes (see STOIIP and GIIP) for a given exploration prospect will allow explorers and commercial analysts to determine whether a prospect is financially viable.

Traditionally, porosity and permeability were determined through the study of drilling samples, analysis of cores obtained from the wellbore, examination of contiguous parts of the reservoir that outcrop at the surface (see e.g. Guerriero et al., 2009, 2011, in references below) and by the technique of formation evaluation using wireline tools passed down the well itself. Modern advances in seismic data acquisition and processing have meant that seismic attributes of subsurface rocks are readily available and can be used to infer physical/sedimentary properties of the rocks themselves.

MINING

Mining is extraction of valuable minerals or other geological materials from the earth usually from an orebody, lode, vein, seam, reef or placer deposits. These deposits form a mineralized package that is of economic interest to the miner.

Surface coal mining

Ores recovered by mining include metals, coal, oil shale, gemstones, limestone, dimension stone, rock salt, potash, gravel, and clay. Mining is required to obtain any material that cannot be grown through agricultural processes, or created artificially in a laboratory or factory. Mining in a wider sense includes extraction of any non-renewable resource such as petroleum, natural gas, or even water.

Mining of stones and metal has been a human activity since pre-historic times. Modern mining processes involve prospecting for ore bodies, analysis of the profit potential of a proposed mine, extraction of the desired materials, and final reclamation of the land after the mine is closed.

Sulfur miner with 90 kg of sulfur carried from the floor of the Ijen Volcano (2015)

Mining operations usually create a negative environmental impact, both during the mining activity and after the mine has closed. Hence, most of the world›s nations have passed regulations to decrease the impact. Work safety has long been a concern as well, and modern practices have significantly improved safety in mines.

Levels of metals recycling are generally low. Unless future end-of-life recycling rates are stepped up, some rare metals may become unavailable for use in a variety of consumer products. Due to the low recycling rates, some landfills now contain higher concentrations of metal than mines themselves.

Prehistoric Mining

Chalcolithic copper mine in Timna Valley, Negev Desert

Since the beginning of civilization, people have used stone, ceramics and, later, metals found close to the Earth›s surface. These were used to make early tools and weapons; for example, high quality flint found in northern France, southern England and Poland was used to create flint tools. Flint mines have been found in chalk areas where seams of the stone were followed underground by shafts and galleries. The mines at Grimes Graves and Krzemionki are especially famous, and like most other flint mines, are Neolithic in origin (ca 4000 BC-ca 3000 BC). Other hard rocks mined or collected for axes included the greenstone of the Langdale axe industry based in the English Lake District.

The oldest-known mine on archaeological record is the Lion Cave in Swaziland, which radiocarbon dating shows to be about 43,000 years old. At this site Paleolithic humans mined hematite to make the red pigment ochre. Mines of a similar age in Hungary are believed to be sites where Neanderthals may have mined flint for weapons and tools.

Ancient Egypt

Ancient Egyptians mined malachite at Maadi. At first, Egyptians used the bright green malachite stones for ornamentations and pottery. Later, between 2613 and 2494 BC, large building projects required expeditions abroad to the area of Wadi Maghareh in order to secure minerals and other resources not available in Egypt itself. Quarries for turquoise and copper were also found at Wadi Hammamat, Tura, Aswan and various other Nubian sites on the Sinai Peninsula and at Timna.

Mining in Egypt occurred in the earliest dynasties. The gold mines of Nubia were among the largest and most extensive of any in Ancient Egypt. These mines are described by the Greek author Diodorus Siculus, who mentions fire-setting as one method used to break down the hard rock holding the gold. One of the complexes is shown in one of the earliest known maps. The miners crushed the ore and ground it to a fine powder before washing the powder for the gold dust.

Ancient Greek and Roman mining

Mining in Europe has a very long history. Examples include the silver mines of Laurium, which helped support the Greek city state of Athens. Although they had over 20,000 slaves working them, their technology was essentially identical to their Bronze Age predecessors. At other mines, such as on the island of Thassos, marble was quarried by the Parians after they arrived in the 7th Century BC. The marble was shipped away and was later found by archaeologists to have been used in buildings including the tomb of Amphipolis. Philip II of Macedon, the father of Alexander the Great, captured the gold mines of Mount Pangeo in 357 BC to fund his military campaigns. He also captured gold mines in Thrace for minting coinage, eventually producing 26 tons per year.

However, it was the Romans who developed large scale mining methods, especially the use of large volumes of water brought to the minehead by numerous aqueducts. The water was used for a variety of purposes, including removing overburden and rock debris, called hydraulic mining, as well as washing comminuted, or crushed, ores and driving simple machinery.

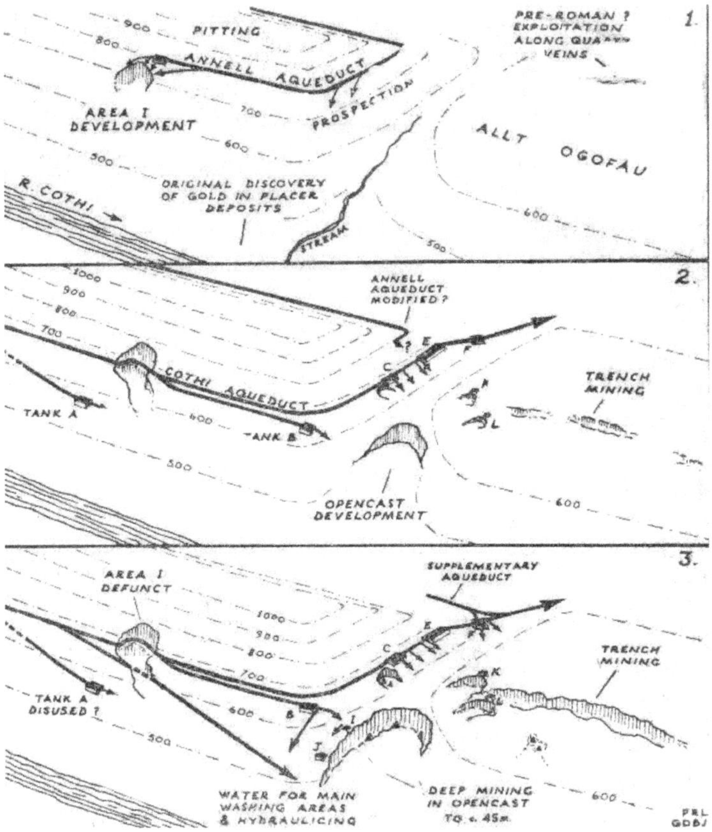

Ancient Roman development of the Dolaucothi Gold Mines, Wales

The Romans used hydraulic mining methods on a large scale to prospect for the veins of ore, especially a now-obsolete form of mining known as hushing. They built numerous aqueducts to supply water to the minehead. There, the water stored in large reservoirs and tanks. When a full tank was opened, the flood of water sluiced away the overburden to expose the bedrock underneath and any gold veins. The rock was then worked upon by fire-setting to heat the rock, which would be quenched with a stream of water. The resulting thermal shock cracked the rock, enabling it to be removed by further streams of water from the overhead tanks. The Roman miners used similar methods to work cassiterite deposits in Cornwall and lead ore in the Pennines.

The methods had been developed by the Romans in Spain in 25 AD to exploit large alluvial gold deposits, the largest site being at Las Medulas, where seven long aqueducts tapped local rivers and sluiced the deposits. Spain was one of the most important mining regions, but all regions of the Roman Empire were exploited. In Great Britain the natives had mined minerals for millennia, but after the Roman conquest, the scale of the operations increased dramatically, as the Romans needed Britannia›s resources, especially gold, silver, tin, and lead.

Roman techniques were not limited to surface mining. They followed the ore veins underground once opencast mining was no longer feasible. At Dolaucothi they stoped out the veins and drove adits through bare rock to drain the stopes. The same adits were also used to ventilate the workings, especially important when fire-setting was used. At other parts of the site, they penetrated the water table and dewatered the mines using several kinds of machines, especially reverse overshot water-wheels. These were used extensively in the copper mines at Rio Tinto in Spain, where one sequence comprised 16 such wheels arranged in pairs, and lifting water about 80 feet (24 m). They were worked as treadmills with miners standing on the top slats. Many examples of such devices have been found in old Roman mines and some examples are now preserved in the British Museum and the National Museum of Wales.

Medieval Europe

Mining and metallurgy in medieval Europe

Agricola, author of *De Re Metallica*

Mining as an industry underwent dramatic changes in medieval Europe. The mining industry in the early Middle Ages was mainly focused on the extraction of copper and iron. Other precious metals were also used, mainly for gilding or coinage. Initially, many metals were obtained through open-pit mining, and ore was primarily extracted from shallow depths, rather than through deep mine shafts. Around the 14th century, the growing use of weapons, armour, stirrups, and horseshoes greatly increased the demand for iron. Medieval knights, for example, were often laden with up to 100 pounds of plate or chain link armour in addition to swords, lances and other weapons. The overwhelming dependency on iron for military purposes spurred iron production and extraction processes.

The silver crisis of 1465 occurred when all mines had reached depths at which the shafts could no longer be pumped dry with the available technology. Although an increased use of bank notes, credit and copper coins during this period did decrease the value of, and dependence on, precious metals, gold and silver still remained vital to the story of medieval mining.

Gallery, 12th to 13th century, Germany

Due to differences in the social structure of society, the increasing extraction of mineral deposits spread from central Europe to England in the mid-sixteenth century. On the continent, mineral deposits belonged to the crown, and this regalian right was stoutly maintained. But in England, royal mining rights were restricted to gold and silver (of which England had virtually no deposits) by a judicial decision of 1568 and a law in 1688. England had iron, zinc, copper, lead, and tin ores. Landlords who owned the base metals and coal under their estates then had a strong inducement to extract these metals or to lease the deposits and collect royalties from mine operators. English, German, and Dutch capital combined to finance extraction and refining. Hundreds of German technicians and skilled workers were brought over; in 1642 a colony of 4,000 foreigners was mining and smelting copper at Keswick in the northwestern mountains.

Use of water power in the form of water mills was extensive. The water mills were employed in crushing ore, raising ore from shafts, and ventilating galleries by powering giant bellows. Black powder was first used in mining in Selmecbánya, Kingdom of Hungary (now Banská Štiavnica, Slovakia) in 1627. Black powder allowed blasting of rock and earth to loosen and reveal ore veins. Blasting was much faster than fire-setting and allowed the mining of previously impenetrable metals and ores. In 1762, the world's first mining academy was established in the same town.

The widespread adoption of agricultural innovations such as the iron plowshare, as well as the growing use of metal as a building material, was also a driving force in the tremendous growth of the iron industry during this period. Inventions like the arrastra were often used by the Spanish to pulverize ore after being mined. This device was powered by animals and used the same principles used for grain threshing.

Much of the knowledge of medieval mining techniques comes from books such as Biringuccio's *De la pirotechnia* and probably most importantly from Georg Agricola›s *De re metallica* (1556). These books detail many different mining methods used in German and Saxon mines. A prime issue in medieval mines, which Agricola explains in detail, was the removal of water from mining shafts. As miners dug deeper to access new veins, flooding became a very real obstacle. The mining industry became dramatically more efficient and prosperous with the invention of mechanical and animal driven pumps.

Classical Philippine civilization

The image of a Maharlika class of the Philippine Society , depicted in Boxer Codex that the Gold used as a form of Jewelry (ca.1400).

Mining in the Philippines began around 1000 BC. The early Filipinos worked various mines of gold, silver, copper and iron. Jewels, gold ingots, chains, calombigas and earrings were handed down from antiquity and inherited from their ancestors. Gold dagger handles, gold dishes, tooth plating, and huge gold ornamets were also used. In Laszlo Legeza›s Tantric elements in pre-Hispanic Philippines Gold Art, he mentioned that gold jewelry of Philippine origin was found in Ancient Egypt. According to Antonio Pigafetta, the people of Mindoro possessed great skill in mixing gold with other metals and gave it a natural and perfect appearance that could deceive even the best of silversmiths. The natives were also known for the jewelries made of other precious stones such as carnelian, agate and pearl. Some outstanding examples of Philippine jewelry included necklaces, belts, armlets and rings placed around the waist.

The Americas

Lead mining in the upper Mississippi River region of the U.S., 1865.

There are ancient, prehistoric copper mines along Lake Superior, and metallic copper was still found there, near the surface, in colonial times.

Indegenous peoples availed themselves of this copper starting at least 5,000 years ago, and copper tools, arrowheads, and other artifacts that were part of an extensive native trade network have been discovered. In addition, obsidian, flint, and other minerals were mined, worked, and traded. Early French explorers who encountered the sites made no use of the metals due to the difficulties of transporting them, but the copper was eventually traded throughout the continent along major river routes.

In the early colonial history of the Americas, native gold and silver was quickly expropriated and sent back to Spain in fleets of gold- and silver-laden galleons, the gold and silver originating mostly from mines in Central and South America. Turquoise dated at 700 A.D. was mined in pre-Columbian America; in the Cerillos Mining District in New Mexico, estimates are that about 15,000 tons of rock had been removed from Mt. Chalchihuitl using stone tools before 1700."

Mining in the United States became prevalent in the 19th century, and the General Mining Act of 1872 was passed to encourage mining of federal lands. As with the California Gold Rush in the mid-19th century, mining for minerals and precious metals, along with ranching, was a driving factor in the Westward Expansion to the Pacific coast. With the exploration of the West, mining camps were established and "expressed a distinctive spirit, an enduring legacy to the new nation;" Gold Rushers would experience the same problems as the Land Rushers of the transient West that preceded them. Aided by railroads, many traveled West for work opportunities in mining. Western cities such as Denver and Sacramento originated as mining towns.

Miners at the Tamarack Mine in Copper Country, Michigan, U.S. in 1905.

When new areas were explored, it was usually the gold (placer and then lode) and then silver that were taken into possession and extracted first. Other metals would often wait for railroads or canals, as coarse gold dust and nuggets do not require smelting and are easy to identify and transport.

Modern Period

In the early 20th century, the gold and silver rush to the western United States also stimulated mining for coal as well as base metals such as copper, lead, and iron. Areas in modern Montana, Utah, Arizona, and later Alaska became predominate suppliers of copper to the world, which was increasingly demanding copper for electrical and households goods. Canada's mining industry grew more slowly than did the United States' due to limitations in transportation, capital, and U.S. competition; Ontario was the major producer of the early 20th century with nickel, copper, and gold.

Meanwhile, Australia experienced the Australian gold rushes and by the 1850s was producing 40% of the world's gold, followed by the establishment of large mines such as the Mount Morgan Mine, which ran for nearly a hundred years, Broken Hill ore deposit (one of the largest zinc-lead ore deposits), and the iron ore mines at Iron Knob. After declines in production, another boom in mining occurred in the 1960s. Now, in the early 21st century, Australia remains a major world mineral producer.

As the 21st century begins, a globalized mining industry of large multinational corporations has arisen. Peak minerals and environmental impacts have also become a concern. Different elements, particularly rare earth minerals, have begun to increase in demand as a result of new technologies.

Mine Development and Lifecycle

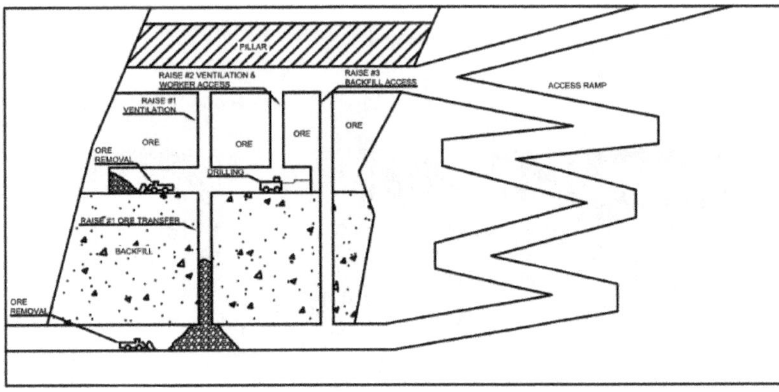

Schematic of a cut and fill mining operation in hard rock.

The process of mining from discovery of an ore body through extraction of minerals and finally to returning the land to its natural state consists of several distinct steps. The first is discovery of the ore body, which is carried out through prospecting or exploration to find and then define the extent, location and value of the ore body. This leads to a mathematical resource estimation to estimate the size and grade of the deposit.

This estimation is used to conduct a pre-feasibility study to determine the theoretical economics of the ore deposit. This identifies, early on, whether further investment in estimation and engineering studies is warranted and identifies key risks and areas for further work. The next step is to conduct a feasibility study to evaluate the financial viability, the technical and financial risks, and the robustness of the project.

This is when the mining company makes the decision whether to develop the mine or to walk away from the project. This includes mine planning to evaluate the economically recoverable portion of the deposit, the metallurgy and ore recoverability, marketability and payability of the ore concentrates, engineering concerns, milling and infrastructure costs, finance and equity requirements, and an analysis of the proposed mine from the initial excavation all the way through to reclamation. The proportion of a deposit that is economically recoverable is dependent on the enrichment factor of the ore in the area.

To gain access to the mineral deposit within an area it is often necessary to mine through or remove waste material which is not of immediate interest to the miner. The total movement of ore and waste constitutes the mining process. Often more waste than ore is mined during the life of a mine, depending on the nature and location of the ore body. Waste removal and placement is a major cost to the mining operator, so a detailed characterization of the waste material forms an essential part of the geological exploration program for a mining operation.

Once the analysis determines a given ore body is worth recovering, development begins to create access to the ore body. The mine buildings and processing plants are built, and any necessary equipment is obtained. The operation of the mine to recover the ore begins and continues as long as the company operating the mine finds it economical to do so. Once all the ore that the mine can produce profitably is recovered, reclamation begins to make the land used by the mine suitable for future use.

Mining Techniques

Underground longwall mining.

Mining techniques can be divided into two common excavation types: surface mining and sub-surface (underground) mining. Today, surface mining is much more common, and produces, for example, 85% of minerals (excluding petroleum and natural gas) in the United States, including 98% of metallic ores.

Targets are divided into two general categories of materials: *placer deposits*, consisting of valuable minerals contained within river gravels, beach sands, and other unconsolidated materials; and *lode deposits*, where valuable minerals are found in veins, in layers, or in mineral grains generally distributed throughout a mass of actual rock. Both types of ore deposit, placer or lode, are mined by both surface and underground methods.

Some mining, including much of the rare earth elements and uranium mining, is done by less-common methods, such as in-situ leaching: this technique involves digging neither at the surface nor underground. The extraction of target minerals by this technique requires that they be soluble, e.g., potash, potassium chloride, sodium chloride, sodium sulfate, which dissolve in water. Some minerals, such as copper minerals and uranium oxide, require acid or carbonate solutions to dissolve.

Surface Mining

Surface mining is done by removing (stripping) surface vegetation, dirt, and, if necessary, layers of bedrock in order to reach buried ore deposits. Techniques of surface mining include: open-pit mining, which is the recovery of materials from an open pit in the ground, quarrying, identical to open-pit mining except that it refers to sand, stone and clay; strip mining, which consists of stripping surface layers off to reveal ore/seams underneath; and mountaintop removal, commonly associated with coal mining, which involves taking the top of a mountain off to reach ore deposits at depth. Most (but not all) placer deposits, because of their shallowly buried nature, are mined by surface methods. Finally, landfill mining involves sites where landfills are excavated and processed.

Garzweiler surface mine, Germany

Underground Mining

Mantrip used for transporting miners within an underground mine

Sub-surface mining consists of digging tunnels or shafts into the earth to reach buried ore deposits. Ore, for processing, and waste rock, for disposal, are brought to the surface through the tunnels and shafts. Sub-surface mining can be classified by the type of access shafts used, the extraction method or the technique used to reach the mineral deposit. Drift mining utilizes horizontal access tunnels, slope mining uses diagonally sloping access shafts, and shaft mining utilizes vertical access shafts. Mining in hard and soft rock formations require different techniques.

Other methods include shrinkage stope mining, which is mining upward, creating a sloping underground room, long wall mining, which is grinding a long ore surface underground, and room and pillar mining, which is removing ore from rooms while leaving pillars in place to support the roof of the room. Room and pillar mining often leads to retreat mining, in which supporting pillars are removed as miners retreat, allowing the room to cave in, thereby loosening more ore. Additional sub-surface mining methods include hard rock mining, which is mining of hard rock (igneous, metamorphic or sedimentary) materials, bore hole mining, drift and fill mining, long hole slope mining, sub level caving, and block caving.

Highwall Mining

Caterpillar Highwall Miner HW300 - Technology Bridging Underground and Open Pit Mining

Highwall mining is another form of surface mining that evolved from auger mining. In Highwall mining, the coal seam is penetrated by a continuous miner propelled by a hydraulic Pushbeam Transfer Mechanism (PTM). A typical cycle includes sumping (launch-pushing forward) and shearing (raising and lowering the cutterhead boom to cut the entire height of the coal seam). As the coal recovery cycle continues, the cutterhead is progressively launched into the coal seam for 19.72 feet (6.01 m). Then, the Pushbeam Transfer Mechanism (PTM) automatically inserts a 19.72-foot (6.01 m) long rectangular Pushbeam (Screw-Conveyor Segment) into the center section of the machine between the Powerhead and the cutterhead. The Pushbeam system can penetrate nearly 1,000 feet (300 m) into the coal seam. One patented Highwall mining system uses augers enclosed inside the Pushbeam that prevent the mined coal from being contaminated by rock debris during the conveyance process. Using a video imaging and/or a gamma ray sensor and/or other Geo-Radar systems like a coal-rock interface detection sensor (CID), the operator can see ahead projection of the seam-rock interface and guide the continuous miner's progress. Highwall mining can produce thousands of tons of coal in contour-strip operations with narrow benches, previously mined areas, trench mine applications and steep-dip seams with controlled water-inflow pump system and/or a gas (inert) venting system.

Machines

The Bagger 288 is a bucket-wheel excavator used in strip mining. It is also the largest land vehicle of all time.

A Bucyrus Erie 2570 dragline and CAT 797 haul truck at the North Antelope Rochelle opencut coal mine

Heavy machinery is used in mining to explore and develop sites, to remove and stockpile overburden, to break and remove rocks of various hardness and toughness, to process the ore, and to carry out reclamation projects after the mine is closed. Bulldozers, drills, explosives and trucks are all necessary for excavating the land. In the case of placer mining, unconsolidated gravel, or alluvium, is fed into machinery consisting of a hopper and a shaking screen or trommel which frees the desired minerals from the waste gravel. The minerals are then concentrated using sluices or jigs.

Large drills are used to sink shafts, excavate stopes, and obtain samples for analysis. Trams are used to transport miners, minerals and waste. Lifts carry miners into and out of mines, and move rock and ore out, and machinery in and out, of underground mines. Huge trucks, shovels and cranes are employed in surface mining to move large quantities of overburden and ore. Processing plants utilize large crushers, mills, reactors, roasters and other equipment to consolidate the mineral-rich material and extract the desired compounds and metals from the ore.

Processing

Once the mineral is extracted, it is often then processed. The science of extractive metallurgy is a specialized area in the science of metallurgy that studies the extraction of valuable metals from their ores, especially through chemical or mechanical means.

Mineral processing (or mineral dressing) is a specialized area in the science of metallurgy that studies the mechanical means of crushing, grinding, and washing that enable the separation (extractive metallurgy) of valuable metals or minerals from their gangue (waste material). Processing of placer ore material consists of gravity-dependent methods of separation, such as sluice boxes. Only minor shaking or washing may be necessary to disaggregate (unclump) the sands or gravels before processing. Processing of ore from a lode mine, whether it is a surface or subsurface mine, requires that the rock ore be crushed and pulverized before extraction of the valuable minerals begins. After lode ore is crushed, recovery of the valuable minerals is done by one, or a combination of several, mechanical and chemical techniques.

Since most metals are present in ores as oxides or sulfides, the metal needs to be reduced to its metallic form. This can be accomplished through chemical means such as smelting or through electrolytic reduction, as in the case of aluminium. Geometallurgy combines the geologic sciences with extractive metallurgy and mining.

Environmental Effects

Environmental issues can include erosion, formation of sinkholes, loss of biodiversity, and contamination of soil, groundwater and surface water by chemicals from mining processes. In some cases, additional forest logging is done in the vicinity of mines to create space for the storage of the created debris and soil. Contamination resulting from leakage of chemicals can also affect the health of the local population if not properly controlled.Extreme examples of pollution from mining activities include coal fires, which can last for years or even decades, producing massive amounts of environmental damage.

Mining companies in most countries are required to follow stringent environmental and rehabilitation codes in order to minimize environmental impact and avoid impacting human health. These codes and regulations all require the common steps of environmental impact assessment, development of environmental management plans, mine closure planning (which must be done before the start of mining operations), and environmental monitoring during operation and after closure. However, in some areas, particularly in the developing world, government regulations may not be well enforced.

Iron hydroxide precipitate stains a stream receiving acid drainage from surface coal mining.

For major mining companies and any company seeking international financing, there are a number of other mechanisms to enforce good environmental standards. These generally relate to financing standards such as the Equator Principles, IFC environmental standards, and criteria for Socially responsible investing. Mining companies have used this oversight from the financial sector to argue for some level of industry self-regulation. In 1992, a Draft Code of Conduct for Transnational Corporations was proposed at the Rio Earth Summit by the UN Centre for Transnational Corporations (UNCTC), but the Business Council for Sustainable Development (BCSD) together with the International Chamber of Commerce (ICC) argued successfully for self-regulation instead.

This was followed by the Global Mining Initiative which was begun by nine of the largest metals and mining companies and which led to the formation of the International Council on Mining and Metals, whose purpose was to act as a catalyst in an effort to improve social and environmental performance in the mining and metals industry internationally. The mining industry has provided funding to various conservation groups, some of which have been working with conservation agendas that are at odds with an emerging acceptance of the rights of indigenous people – particularly the right to make land-use decisions.

Certification of mines with good practices occurs through the International Organization for Standardization (ISO). For example, ISO 9000 and ISO 14001, which certify an auditable environmental management system, involve short inspections, although they have been accused of lacking rigor. Certification is also available through Ceres› Global Reporting Initiative, but these reports are voluntary and unverified. Miscellaneous other certification programs exist for various projects, typically through nonprofit groups. The purpose of a 2012 EPS PEAKS paper was to provide evidence on policies managing ecological costs and maximise socio-economic benefits of mining using host country regulatory initiatives. It found existing literature suggesting donors encourage developing countries to:

- Make the environment-poverty link and introduce cutting-edge wealth measures and natural capital accounts.
- Reform old taxes in line with more recent financial innovation, engage directly with the companies, enacting land use and impact assessments, and incorporate specialised support and standards agencies.
- Set in play transparency and community participation initiatives using the wealth accrued.

Waste

Ore mills generate large amounts of waste, called tailings. For example, 99 tons of waste are generated per ton of copper, with even higher ratios in gold mining - because only 5.3 g of gold is extracted per ton of ore, a ton of gold produces 200,000 tons of tailings. (As time goes on and richer deposits are exhausted - and technology improves to permit - this number is going down to .5 g and less.) These tailings can be toxic. Tailings, which are usually produced as a slurry, are most commonly dumped into ponds made from naturally existing valleys. These ponds are secured by impoundments (dams or embankment dams). In 2000 it was estimated that 3,500 tailings impoundments existed, and that every year, 2 to 5 major failures and 35 minor failures occurred; for example, in the Marcopper mining disaster at least 2 million tons of tailings were released into a local river. Subaqueous tailings disposal is another option. The mining industry has argued that submarine tailings disposal (STD), which disposes of tailings in the sea, is ideal because it avoids the risks of tailings ponds; although the practice is illegal in the United States and Canada, it is used in the developing world.

The waste is classified as either sterile or mineralised, with acid generating potential, and the movement and storage of this material forms a major part of the mine planning process. When the mineralised package is determined by an economic cut-off, the near-grade mineralised waste is usually dumped separately with view to later treatment should market conditions change and it becomes economically viable. Civil engineering design parameters are used in the design of the waste dumps, and special conditions apply to high-rainfall areas and to seismically active areas. Waste dump designs must meet all regulatory requirements of the country in whose jurisdiction the mine is located. It is also common practice to rehabilitate dumps to an internationally acceptable standard, which in some cases means that higher standards than the local regulatory standard are applied.

Renewable Energy and Mining

Many mining sites are remote and not connected to the grid. Electricity is typically generated with diesel generators. Due to high transportation cost and theft during transportation the cost for generating electricity is normally high. Renewable energy applications are becoming an alternative or amendment. Both solar and wind power plants can contribute in saving diesel costs at mining sites. Renewable energy applications have been built at mining sites. Cost savings can reach up to 70%.

Mining Industry

Mining exists in many countries. London is known as the capital of global "mining houses" such as Rio Tinto Group, BHP Billiton, and Anglo American PLC. The US mining industry is also large, but it is dominated by the coal and other nonmetal minerals (e.g., rock and sand), and various regulations have worked to reduce the significance of mining in the United States. In 2007 the total market capitalization of mining companies was reported at US$962 billion, which compares to a total global market cap of publicly traded companies of about US$50 trillion in 2007. In 2002, Chile and Peru were reportedly the major mining countries of South America. The mineral industry of Africa includes the mining of various minerals; it produces relatively little of the industrial metals copper, lead, and zinc, but according to one estimate has as a percent of world reserves 40% of gold, 60% of cobalt, and 90% of the world's platinum group metals. Mining in India is a significant part of that country's economy. In the developed world, mining in Australia, with BHP Billiton founded and headquartered in the country, and mining in Canada are particularly significant. For rare earth minerals mining, China reportedly controlled 95% of production in 2013.

The Bingham Canyon Mine of Rio Tinto's subsidiary, Kennecott Utah Copper.

While exploration and mining can be conducted by individual entrepreneurs or small businesses, most modern-day mines are large enterprises requiring large amounts of capital to establish. Consequently, the mining sector of the industry is dominated by large, often multinational, companies, most of them publicly listed. It can be argued that what is referred to as the ‹mining industry› is actually two sectors, one specializing in exploration for new resources and the other in mining those resources. The exploration sector is typically made up of individuals and small mineral resource companies, called juniors, which are dependent on venture capital. The mining sector is made up of large multinational companies that are sustained by production from their mining operations. Various other industries such as equipment manufacture, environmental testing, and metallurgy analysis rely on, and support, the mining industry throughout the world. Canadian stock exchanges have a particular focus on mining companies, particularly junior exploration companies through Toronto›s TSX Venture Exchange; Canadian companies raise capital on these exchanges and then invest the money in exploration globally. Some have argued that below juniors there exists a substantial sector of illegitimate companies primarily focused on manipulating stock prices.

Mining operations can be grouped into five major categories in terms of their respective resources. These are oil and gas extraction, coal mining, metal ore mining, nonmetallic mineral mining and quarrying, and mining support activities. Of all of these categories, oil and gas extraction remains one of the largest in terms of its global economic importance. Prospecting potential mining sites, a vital area of concern for the mining industry, is now done using sophisticated new technologies such as seismic prospecting and remote-sensing satellites. Mining is heavily affected by the prices of the commodity minerals, which are often volatile. The 2000s commodities boom (“commodities supercycle) increased the prices of commodities, driving aggressive mining. In addition, the price of gold increased dramatically in the 2000s, which increased gold mining; for example, one study found that conversion of forest in the Amazon increased six-fold from the period 2003–2006 (292 ha/yr) to the period 2006–2009 (1,915 ha/yr), largely due to artisanal mining.

Corporate Classifications

Mining companies can be classified based on their size and financial capabilities:

- **Major** companies are considered to have an adjusted annual mining-related revenue of more than US$500 million, with the financial capability to develop a major mine on its own.
- **Intermediate** companies have at least $50 million in annual revenue but less than $500 million.
- **Junior** companies rely on equity financing as their principal means of funding exploration. Juniors are mainly pure exploration companies, but may also produce minimally, and do not have a revenue exceeding US$50 million.

Regulation and Governance

New regulations and a process of legislative reforms aim to improve the harmonization and stability of the mining sector in mineral-rich countries. New legislation for mining industry in African countries still appears to be an issue, but has the potential to be solved, when a consensus is reached on the best approach. By the beginning of the 21st century the booming and increasingly complex mining sector in mineral-rich countries was providing only slight benefits to local communities, especially in given the sustainability issues. Increasing debate and influence by NGOs and local communities called for a new approahes which would also include disadvantaged communities, and work towards sustainable development even after mine closure (including transparency and revenue management). By the early 2000s, community development issues and resettlements became mainstream concerns in World Bank mining projects. Mining-industry expansion after mineral prices increased in 2003 and also potential fiscal revenues in those countries created an omission in the other economic sectors in terms of finances and development. Furthermore, this highlighted regional and local demand for mining revenues and an inability of sub-national governments to effectively use the revenues. The Fraser Institute (a Canadian think tank) has highlighted the environmental protection laws in developing countries, as well as voluntary efforts by mining companies to improve their environmental impact.

In 2007 the Extractive Industries Transparency Initiative (EITI) was mainstreamedin all countries cooperating with the World Bank in mining industry reform. The EITI operates and was implemented with the support of the EITI multi-donor trust fund, managed by the World Bank. The Extractive Industries Transparency Initiative (EITI) aims to increase transparency in transactions between governments and companies in extractive industries by monitoring the revenues and benefits between industries and recipient governments. The entrance process is voluntary for each country and is monitored by multiple stakeholders including governments, private companies and civil society representatives, responsible for disclosure and dissemination of the reconciliation report; however, the competitive disadvantage of company-by company public report is for some of the businesses in Ghana at least, the main constraint. Therefore, the outcome assessment in terms of failure or success of the new EITI regulation does not only rest on the government›s shoulders but also on civil society and companies.

On the other hand implementation has issues; inclusion or exclusion of artisanal mining and small-scale mining (ASM) from the EITI and how to deal with "non-cash" payments made by companies to subnational governments. Furthermore, the disproportionate revenues the mining industry can bring to the comparatively small number of people that it employs, causes other problems, like a lack of investment in other less lucrative sectors, leading to swings in government revenuebecause of volatility in the oil markets. Artisanal mining is clearly an issue in EITI Countries such as the Central African Republic, D.R. Congo, Guinea, Liberia and Sierra Leone – i.e. almost half of the mining countries implementing the EITI.Among other things, limited scope of the EITI involving disparity in terms of knowledge of the industry and negotiation skills, thus far flexibility of the policy (e.g. liberty of the countries to expand beyond the minimum requirements and adapt it to their needs), creates another risk of unsuccessful implementation. Public awareness increase, where government should act as a bridge between public and initiative for a successful outcome of the policy is an important element to be considered.

World Bank

The World Bank has been involved in mining since 1955, mainly through grants from its International Bank for Reconstruction and Development, with the Bank›s Multilateral Investment Guarantee Agency offering political risk insurance. Between 1955 and 1990 it provided about $2 billion to fifty mining projects, broadly categorized as reform and rehabilitation, greenfield mine construction, mineral processing, technical assistance, and engineering. These projects have been criticized, particularly the Ferro Carajas project of Brazil, begun in 1981. The World Bank established mining codes intended to increase foreign investment; in 1988 it solicited feedback from 45 mining companies on how to increase their involvement.[20]

In 1992 the World Bank began to push for privatization of government-owned mining companies with a new set of codes, beginning with its report *The Strategy for African Mining*. In 1997, Latin America's largest miner Companhia Vale do Rio Doce (CVRD) was privatized. These and other developments such as the Philippines 1995 Mining Act led the bank to publish a third report (*Assistance for Minerals Sector Development and Reform in Member Countries*) which endorsed mandatory environment impact assessments and attention to the concerns of the local population. The codes based on this report are influential in the legislation of developing nations. The new codes are intended to encourage development through tax holidays, zero custom duties, reduced income taxes, and related measures.[22] The results of these codes were analyzed by a group from the University of Quebec, which concluded that the codes promote foreign investment but "fall very short of permitting sustainable development". The observed negative correlation between natural resources and economic development is known as the resource curse.

Safety

Mining transport in Devnya, Bulgaria.

Mine safety

Safety has long been a concern in the mining business, especially in sub-surface mining. The Courrières mine disaster, Europe's worst mining accident, involved the death of 1,099 miners in Northern France on March 10, 1906. This disaster was surpassed only by the Benxihu Colliery accident in China on April 26, 1942, which killed 1,549 miners. While mining today is substantially safer than it was in previous decades, mining accidents still occur. Government figures indicate that 5,000 Chinese miners die in accidents each year, while other reports have suggested a figure as high as 20,000. Mining accidents continue worldwide, including accidents causing dozens of fatalities at a time such as the 2007 Ulyanovskaya Mine disaster in Russia, the 2009 Heilongjiang mine explosion in China, and the 2010 Upper Big Branch Mine disaster in the United States.

There are numerous occupational hazards associated with mining, including exposure to rockdust which can lead to diseases such as silicosis, asbestosis, and pneumoconiosis. Gases in the mine can lead to asphyxiation and could also be ignited. Mining equipment can generate considerable noise, putting workers at risk for hearing loss. Cave-ins, rock falls, and exposure to excess heat are also known hazards.

Proper ventilation, hearing protection, and spraying equipment with water are important safety practices in mines.

Records

Chuquicamata, Chile, site of the largest circumference and second deepest open pit copper mine in the world.

As of 2008, the deepest mine in the world is TauTona in Carletonville, South Africa at 3.9 kilometres (2.4 mi), replacing the neighboring Savuka Mine in the North West Province of South Africa at 3,774 metres (12,382 ft). East Rand Mine in Boksburg, South Africa briefly held the record at 3,585 metres (11,762 ft), and the first mine declared the deepest in the world was also TauTona when it was at 3,581 metres (11,749 ft).

The Moab Khutsong gold mine in North West Province (South Africa) has the world's longest winding steel wire rope, able to lower workers to 3,054 metres (10,020 ft) in one uninterrupted four-minute journey.

The deepest mine in Europe is the 16th shaft of the uranium mines in Příbram, Czech Republic at 1,838 metres (6,030 ft), second is Bergwerk Saar in Saarland, Germany at 1,750 metres (5,740 ft).

The deepest open-pit mine in the world is Bingham Canyon Mine in Bingham Canyon, Utah, United States at over 1,200 metres (3,900 ft). The largest and second deepest open-pit copper mine in the world is Chuquicamata in Chuquicamata, Chile at 900 metres (3,000 ft), 443,000 tons of copper and 20,000 tons of molybdenum produced annually.

The deepest open-pit mine with respect to sea level is Tagebau Hambach in Germany, where the base of the pit is 293 metres (961 ft) below sea level.

The largest underground mine is Kiirunavaara Mine in Kiruna, Sweden. With 450 kilometres (280 mi) of roads, 40 million tonnes of ore produced yearly, and a depth of 1,270 metres (4,170 ft), it is also one of the most modern underground mines. The deepest borehole in the world is Kola Superdeep Borehole at 12,262 metres (40,230 ft). This, however, is not a matter of mining but rather related to scientific drilling.

Metal Reserves and Recycling

During the 20th century, the variety of metals used in society grew rapidly. Today, the development of major nations such as China and India and advances in technologies are fueling an ever-greater demand. The result is that metal mining activities are expanding and more and more of the world's metal stocks are above ground in use rather than below ground as unused reserves. An example is the in-use stock of copper. Between 1932 and 1999, copper in use in the USA rose from 73 kilograms (161 lb) to 238 kilograms (525 lb) per person.

95% of the energy used to make aluminium from bauxite ore is saved by using recycled material. However, levels of metals recycling are generally low. In 2010, the International Resource Panel, hosted by the United Nations Environment Programme (UNEP), published reports on metal stocks that exist within society and their recycling rates.

The report›s authors observed that the metal stocks in society can serve as huge mines above ground. However, they warned that the recycling rates of some rare metals used in applications such as mobile phones, battery packs for hybrid cars, and fuel cells are so low that unless future end-of-life recycling rates are dramatically stepped up these critical metals will become unavailable for use in modern technology.

As recycling rates are low and so much metal has already been extracted, some landfills now contain a higher concentrations of metal than mines themselves. This is especially true of aluminium, used in cans, and precious metals, found in discarded electronics. Furthermore, waste after 15 years has still not broken down, so less processing would be required when compared to mining ores. A study undertaken by Cranfield University has found £360 million of metals could be mined from just 4 landfill sites. There is also up to 20MJ/kg of energy in waste, potentially making the re-extraction more profitable.

However, although the first landfill mine opened in Tel Aviv, Israel in 1953, little work has followed due to the abundance of accessible ores.

REFERENCES

1. Hartman, Howard L. SME Mining Engineering Handbook, Society for Mining, Metallurgy, and Exploration Inc, 1992, p3.
2. Swaziland Natural Trust Commission, "Cultural Resources - Malolotja Archaeology, Lion Cavern, Retrieved August 27, 2007, .
3. Peace Parks Foundation, "Major Features: Cultural Importance." Republic of South Africa: Author. Retrieved August 27, 2007, .
4. ASA - October 1996: Mining and Religion in Ancient Man. www2.asa3.org. Retrieved 2015-06-11.
5. Shaw, I. (2000). The Oxford History of Ancient Egypt. New York: Oxford University Press, pp. 57–59.
6. Shaw, I. (2000). The Oxford History of Ancient Egypt. New York: Oxford University Press, p. 108.
7. MINING GREECE ANCIENT MINES. www.miningreece.com. Retrieved 2015-06-11.
8. MINING GREECE ANCIENT QUARRIES IN THASSOS. www.miningreece.com. Retrieved 2015-06-11.
9. MINING GREECE THE GOLDMINES OF ALEXANDER THE GREAT. www.miningreece.com. Retrieved 2015-06-11.
10. The Independent, 20 Jan. 2007: ‹›The end of a Celtic tradition: the last gold miner in Wales››. News.independent.co.uk. 2007-01-20. Archived from the original on July 6, 2008. Retrieved 2013-06-22.
11. The Romans in Britain: mining Archived July 20, 2010, at the Wayback Machine.
12. A culture of Improvement. Robert Friedel. MIT Press. 2007. Pg.81
13. Chapter 7: Medieval Silver and Gold. Mygeologypage.ucdavis.edu. Retrieved 2013-06-22.
14. Heaton Herbert (1948) Economic History of Europe. A Harper International Edition. Fifth printing. February 1968. P.316
15. Heiss, A.G. & Oeggl, K. (2008). Analysis of the fuel wood used in Late Bronze Age and Early Iron Age copper mining sites of the Schwaz and Brixlegg area (Tyrol, Austria). Vegetation History and Archaeobotany 17(2):211-221, Springer Berlin / Heidelberg, .
16. The use of Firesetting in the Granite Quarries of South India Paul T. Craddock The Bulletin of the Peak District Mines Historical Society, Vol. 13 Number 1. 1996
17. "The Spanish Tradition in Gold and Silver Mining." Otis E. Young Arizona and the West, Vol. 7, No. 4 (Winter, 1965), pp. 299-314 (Journal of the Southwest)

Stable URL: http://www.jstor.org/stable/40167137.

18. Ancient Philippine Civilization. Accessed January 7, 2013.(archived from the original on 2007-12-01).
19. Lankton, L. (1991). Cradle to Grave: Life, Work, and Death at the Lake Superior Copper Mines. New York: Oxford University Press, p. 5-6.
20. West, G.A. (1970). Copper: its mining and use by the aborigines of the Lake Superior Region. Westport, Conn: Greenwood Press.
21. Ricard, T. A. (1932), A History of American Mining, McGraw-Hill Book Company.
22. Vaden, H.E. & Prevost. G. (2002). Politics of Latin America: The Power Game. New York: Oxford University Press, p. 34.
23. Maynard, S.R., Lisenbee, A.L. & Rogers, J. (2002). Preliminary Geologic Map of the Picture Rock 7.5 - Minute Quadrangle Sante Fe County, Central New Mexico. New Mexico Bureau of Geology and Mineral Resources, Open-File Report DM-49.
24. The Cerrillos Hills Park Coalition, (2000). Cerrillos Hills Historic Park Vision Statement. Public documents: Author. Retrieved August 27, 2007, . Archived August 1, 2012, at the Wayback Machine.
25. McClure R, Schneider A. The General Mining Act of 1872 has left a legacy of riches and ruin. Seattle PI.
26. Boorstin, D.J. (1965). The Americans: The National Experience. New York: Vintage Books, pp. 78-81.
27. Miller C. (2013). Atlas of US and Canadian Environmental History, p64. Taylor & Francis.
28. History of Australia›s Minerals Industry. Australian Atlas of Minerals Processing, Mines, and & Processing Centres.
29. Hartmann HL. Introductory Mining Engineering, p. 11. First chapter.
30. In Situ Leach Mining (ISL) of Uranium. World-nuclear.org. Retrieved 2013-06-22.
31. Mining, Quarrying & Prospecting: The Difference between Mining, Quarrying & Prospecting. mqp-geotek.blogspot.co.uk. Retrieved 2015-06-11.
32. Landfill Mining Landfill Mining, Preserving Resources through Integrated Sustainable Management of Waste, Technical Brief from the World Resource Foundation
33. Logging of forests and debris dumping. Ngm.nationalgeographic.com. Retrieved 2013-06-22.
34. Larmer, Brook (January 2009). The Real Price of Gold. National Geographic.
35. Moody R. (2007). Rocks and Hard Places. Zed Books.
36. Abrahams D. (2005). Regulations for Corporations: A historical account of TNC regulation, p. 6. UNRISD.
37. Chapin, Mac (2004-10-15). A Challenge to Conservationists: Can we protect

natural habitats without abusing the people who live in them?. World Watch Magazine. 6. 17. Retrieved 2010-02-18.

38. Bloom, M.J. & Denison, M. (2012) Environmental management for extractives, PROFESSIONAL EVIDENCE AND APPLIED KNOWLEDGE SERVICES http://partnerplatform.org/?zl177g4a
39. Consequences of over Exploitation of Mineral Resources. www.preservearticles.com. Retrieved 2015-06-11. |first1= missing |last1= in Authors list (help)
40. What is the Cost of Mining Gold? - Visual Capitalist. Visual Capitalist. Retrieved 2015-06-11.
41. US EPA. (1994). Technical Report: Design and Evaluation of Tailings Dams.
42. TE Martin, MP Davies. (2000). Trends in the stewardship of tailings dams.
43. Coumans C. (2002). Mining's Problem with Waste. MiningWatch Canada.
44. http://www.th-energy.net/english/platform-renewable-energy-and-mining/database-solar-wind-power-plants/
45. http://www.mining.com/web/at-mining-sites-renewable-energy-systems-are-up-to-70-percent-less-expensive-than-diesel-power/
46. MacDonald A. (2002). Industry in Transition: A Profile of the North American Mining Sector. Free full-text.
47. Reuters. Global stock values top $50 trln: industry data.
48. Environmental Effects of Foreign Investment Versus Domestic Investment in the Mining Sector In Latin-America. OECD.
49. Mining in Africa - Overview. Mbendi.
50. China›s Continuing Monopoly Over Rare Earth Minerals. U.S. News & World Report.
51. United States Bureau of Labor http://www.bls.gov/oco/cg/cgs004.htm#nature
52. Swenson JJ, Carter CE, Domec J-C, Delgado CI (2011) Gold Mining in the Peruvian Amazon: Global Prices, Deforestation, and Mercury Imports. PLoS ONE 6(4): e18875. doi:10.1371/journal.pone.0018875. Lay summary: Amazon Gold Fever Comes with a High Environmental Cost.
53. Metals Economics Group World Exploration Trends Report (PDF). Metals Economics Group Inc. Retrieved 2009-05-05.
54. Cambell, Bonnie (2008). Regulation & Legitimacy in the Mining Industry in Africa: Where does (PDF). Review of African Political Economy. 35 (3): 367–389. doi:10.1080/03056240802410984. Retrieved 7 April 2011.
55. The World Bank. ces.worldbank.org/INTOGMC/Resources/336099-1288881181404/7530465-1288881207444/eifd19_mining_sector_reform.pdf The World Bank›s Evolutionary Approach to Mining Sector Reform Check |url= value (help) (PDF). Retrieved 4 April 2011.
56. Do Canadian mining companies operating abroad face weaker environmental regulations?. MiningFacts.org. Fraser Institute.

57. Extractive Industries Transparency Initiative. Governance Structure. Retrieved 4 April 2011.
58. Business and Human Right Resource Centre (2009). Principles: Extractive Industries Transparency Initiative (EITI). Retrieved 6 April 2011.
59. The Ghanaian Journal. At the Fifth EITI Global Conference. Retrieved 3 April 2011.
60. Extractive Industries Transparency Initiative. Report of 5th EITI Global Conference in Paris, 2011 (PDF). Retrieved 4 April 2011.
61. World Bank's Oil, Gas and Mining Policy and Operations Unit (COCPO). Advancing the EITI in the Mining Sector: Implementation Issues (PDF). Retrieved 6 April 2011.
62. Revenue Watch Institute 2010. Promoting Transparency in the Extractive Sectors: An EITI Training for Tanzania Legislators. Archived from the original on July 20, 2011. Retrieved 6 April 2011.
63. For an overview of the Bank and mining, see Mining, Sustainability and Risk:World Bank Group Experiences.
64. See the 1995 World Development 23(3) pp. 385-400.
65. GRAMA. (2003). The Challenges of Development, Mining Codes in Africa And Corporate Responsibility. In: International and Comparative Mineral Law and Policy: Trends and Prospects. Summarized in the African Mining Codes Questioned. ArchivedJanuary 1, 2016, at the Wayback Machine.
66. Marcel Barrois (in French). Le Monde. March 10, 2006.
67. Where The Coal Is Stained With Blood. TIME. March 2, 2007
68. TauTona, Anglo Gold - Mining Technology. SPG Media Group PLC. 2009-01-01. Retrieved 2009-03-02.
69. Naidoo, Brindaveni (2006-12-15). TauTona to take ‹deepest mine› accolade. Creamer Media›s Mining Weekly Online. Archived from the original on 2007-08-19. Retrieved 2007-07-19.
70. Mining and minerals in South Africa. SouthAfrica.info. 8 August 2012. Retrieved 13 August 2012.
71. Mineral deposits: from their origin to their environmental impacts. Taylor & Francis.
72. Chuquicamata | MINING.com. Retrieved 2015-06-11.
73. Chuquicamata›s Life Underground Will Cost a Fortune, but is Likely to Pay Off for Codelco | Copper Investing News. 2015-04-06. Archived from the original on April 6, 2015. Retrieved 2015-06-11.
74. The TEX Report Ltd.. www.texreport.co.jp. Retrieved 2015-06-11.
75. The Recycling Rates of Metals: A Status Report 2010, International Resource Panel, United Nations Environment Programme
76. Tread lightly: Aluminium attack Carolyn Fry, Guardian.co.uk, 22 February 2008.

77. Metal Stocks in Society: Scientific Synthesis 2010, International Resource Panel, United Nations Environment Programme

78. http://www.macfarlanes.com/media/1467/landfill-mining-new-opportunities-ahead.pdf (PDF). www.macfarlanes.com. Retrieved 2015-06-11. External link in |title= (help)

79. Landfill Mining. www.enviroalternatives.com. Retrieved 2015-06-11.

80. Study shows around £360 million of metals could be mined from just four landfill sites. www.rebnews.com. Retrieved 2015-06-11.

81. http://www.elfm.eu/Uploads/ELFM/FILE_79F81D49-34DC-4B5B-9BFC-5E5101CE7520.PDF (PDF). www.elfm.eu. Retrieved 2015-06-11. External link in |title= (help)

82. Assessing the opportunities of landfill mining - Research database - University of Groningen. www.rug.nl. Retrieved 2015-06-11.

Index